P9-CAM-671

BORINGOLOGY

77 nerdy, bizarre and bewildering scientific projects that just might change the world

BORINGOLOGY

77 nerdy, bizarre and bewildering scientific projects that just might change the world

Roger Dobson

Marshall Cavendish
Editions

For my wife Alex, our children, Jessica, Ben, Lizzie and James,
and Mum & Dad

Copyright © 2008 Marshall Cavendish Limited

First published in 2008 by

Marshall Cavendish Limited
5th floor
32–38 Saffron Hill
London ECIN 8FH
United Kingdom
Tel: +44 (0)20 7421 8120
Fax: +44 (0)20 7421 8121
sales@marshallcavendish.co.uk
www.marshallcavendish.co.uk

All rights reserved

Without limiting the rights under copyright reserved above, no part of this
publication may be reproduced, stored in a retrieval system or transmitted in
any form or by any means including photocopying, electronic, mechanical,
recording or otherwise, without the prior written permission of the rights
holders, application for which must be made to the publisher.

A CIP record for this book is available from the British Library

ISBN-13 978-1-905736-15-7
ISBN-10 1-905736-15-0

Printed and bound in Great Britain by
TJ International Ltd, Padstow, Cornwall

Disclaimer
The projects, research and findings discussed in this book are intended solely for
debate. No statement made here constitutes medical or any other kind of advice
or recommendation of any course of action or inaction. No responsibility for any
consequences, however caused, of any acts or omissions arising from any part of
this text will be accepted by the author or Marshall Cavendish Limited.

CONTENTS

FOREWORD

Research is to see what everybody else has seen, and to think
what nobody else has thought ...

Albert Szent-Györgi (1893–1986),
Hungarian Nobel Prize-winning biochemist

AT FIRST GLANCE, THERE DOESN'T seem a lot of point to collecting old
toenails.

Nor does there seem too much of a future in counting how many
times passengers yawn on trains, or watching cabbages grow from an
aeroplane, or measuring and comparing the size of the genitalia of polar
bears, or talking to penguins.

But while research projects like these may seem unusual, bizarre,
mundane, even boring, the answers they provide may have far-reaching
implications.

Experiments with urine on cucumbers, for example, could help
solve some of the world's famine problems, while analysing dandruff
might help put more criminals away, and counting worms on golf
courses just might make millions of golfers happier.

This may be the unglamorous coalface of academic research, an
area rarely illuminated by the media spotlight, or distracted by the
glitter of scientific accolades, but it's home territory for thousands of
researchers who toil to find answers to some of the most unlikely
questions.

In some cases the findings have important ramifications for mankind.
Could the genitalia of polar bears be used as a barometer of global
warming, for example? And could inducing sexual fetishes in birds lead
to better treatment for sexual problems in men and women? And might

measuring how much the female breast bounces lead to better support-wear for millions of women? And could those toenails be a new way to predict heart attacks?

Other researchers have spent time finding answers for some of life's more unusual questions, such as how long does it really take to boil an egg, or how much force is needed to crack an egg, or how long does it take an overturned tortoise to right itself?

Others have gone where few would venture, like the researchers who analyse the contents of nappies, or smell used T-shirts, or sniff men's shirts after they have eaten meat.

But what they all have in common is enthusiasm. Whether it's boring, bizarre or just off the wall, all the researchers set out to find answers to questions they see as important enough to spend time and energy on.

In some cases they may not get the results they want. They may never live to see their findings have any use, and they may spend months, years even, going up blind alleys.

But someone has to do it, and there's always a silver lining, or, as the old laboratory saying goes, "Every experiment proves something. If it doesn't prove what you wanted it to prove, it proves something else."

1

DIRTY NAPPIES, URINE, BODY ODOUR, SHIRT SMELLERS, AND 100,000 OLD TOENAILS

Looking into dirty nappies

TAKING SPOONFULS OF DEBRIS FROM the nappies of babies and looking for bugs may not be the world's most glamorous research job, but it could be vital.

To the untrained eye, and nose, it may reveal little more than clues to what the young providers had eaten, but to the expert, it's a treasure trove.

We may know that baby gets her eyes from Mum, her hair from Dad and her nose from Grandma, but thanks to this research, we may also know where the bacteria in her gut comes from.

And that's important because the contents of the gut play a key role in disease and in digestion. The bugs help extract more goodness from food, keep the immune system healthy, and protect against harmful bugs.

The samples that the researchers regularly collect in the first year of the lives of babies, plus similar material from their mums and dads, are giving a unique insight into the development of the human gut, and just how it helps us to digest food, mitigate disease, regulate fat storage, and even promote the growth of blood vessels

It's known that there are more than four hundred species of bacteria in the intestines, and the average adult human body has ten times more of these cells than human cells.

These multiplying microbes serve numerous purposes, including protecting against harmful pathogens and aiding digestion. But exactly what each does, and how they develop, is largely unknown.

What is known is that immediately before birth, the foetal intestinal tract is sterile. But at birth, the guts of babies rapidly become colonized with microbes from their environment, the birth canal, their mothers' breasts, and even the touch of a sibling or parent.

Within days, a thriving microbial community is established.

At the Howard Hughes Medical Institute and Stanford University, researchers have set out to track the evolution of the gut ecosystems in fourteen breastfed babies.

Immediately after birth and at set times during the first year, stool samples are taken from the newborns using a spoon designed to scoop out 300 mg of material from the nappy. An average of 26 samples a year are obtained from each child, plus reference samples supplied by some of the mums and dads.

By analysing the contents of the samples, researchers are able to watch how the bacteria develop in the gut.

In the research, Dr Chana Palmer and the team collected stool samples from the babies and their parents at several intervals over each baby's first year. DNA from the samples was then spread on a glass chip dotted with known bacterial DNA. Samples whose DNA sequence matched any bacterial sequence on the chip latched on to those spots and were tallied by a computer.

Hundreds of different species of bacteria were found to inhabit an infant's gastrointestinal tract, and each baby had a different mix. The fraternal twins in the study showed the most similarity, suggesting that genetics and environment work together to shape the gut population in a reproducible way. By year one, all the infants had a general profile close to that of an adult.

"It has been recognized for nearly a century that human beings are inhabited by a remarkably dense and diverse microbial ecosystem, yet we are only now just beginning to understand and appreciate the many roles that these microbes play in human health and development. Knowing the composition of this ecosystem is a crucial step toward understanding its roles," says Palmer.

"It almost doesn't matter where you start off because we all end up in the same place," she goes on. "There are some bacteria that are really well suited for your gut and they're going to win no matter what."

Whether bacterial flora are a function of genetics or the environment or both remains to be tested, says Palmer, who likens the process to gardening. "What comes up depends both on what seeds were sown and which are best suited to the particular soil and climate."

She adds, "We found that the composition and temporal patterns of the microbial communities varied widely from baby to baby, supporting a broader definition of healthy colonization than previously recognized."

The researchers say that identifying the environmental and genetic factors that determine the distinctive characteristics of each individual's microbiota, and determining whether and how these individual specific features affect health, will be important goals for future investigations.

How urine could save the world from famine

THE CUCUMBERS BEING GROWN IN a remote part of Finland are no ordinary crop.

Not for them the conventional types of commercial fertilizers or manures. What's making them grow is urine, gallons of the stuff.

Urine from kindergartens, cafés and private homes is being collected, analysed and then used to nourish the cucumbers by researchers at the University of Kuopio, who then watch how well and how fast they grow.

Urine, as researchers point out, is rich in plant nutrients. The human kidney is the main excretion organ and, as a result, urine contains most of the nutrients present in human food that have not been used by the body for new cell growth or energy consumption.

Human urine is a complex water solution containing nutrients as highly diluted compounds. Sodium chloride and urea are the main compounds, although urine also contains, for example, potassium, calcium, sulphate and phosphorus. Around 50 per cent of the phosphorus in municipal waste water originates from urine, and it's been estimated, by researchers at the University of Kalyani in India, that in an average year the urine of one person produces about 4.6 kg of nitrogen, 0.4 kg of phosphorus and 11 kg of potassium.

Urea, one of the main components present in human urine, just happens also to be one of the most important industrial nitrogen fertilizers. New urea-ammonia fertilizer plants have been built in a number of countries in recent years, including India.

Cucumbers were chosen as the test crop for the research because they are grown worldwide, and they are often eaten without cooking. Because they are mostly eaten raw, the research would need to show that there was no potentially harmful contamination arising from the use of urine, and that there was no adverse effect on flavour.

Cucumbers of the Adam (*Bejo zaden*) variety were seeded in a greenhouse and the young seedlings were then planted outdoors into banks. Some of the rows were fertilized with urine and others with a conventional mineral fertilizer.

Cucumbers reaching a length of 10 cm were then harvested and compared. The taste was assessed by a panel of twenty men and women trained in the skill of tasting.

The results show that the growth of both sets of cucumbers was the same, but the total yield was actually greater in those that had been fed urine.

Analysis showed that none of the crop contained any nasty bugs – coliforms, enterococci, coliphages and clostridia – and the tasting results show that the tasters did not prefer any particular cucumber samples, and all the cucumbers were rated as being good tasting and good looking with a nice texture.

"The results show clearly that recently formed urine could serve as a valuable fertilizer for cucumbers, and these vegetables could be eaten without cooking or used for fermentation," say the researchers. "If the amounts of nutrients are correct, urine could totally replace commercial fertilizer, as equal yields are obtained with urine as with normal fertilizer methods."

The researchers say that the implications of the research, especially for the developing world, are immense. Human urine, they point out, is a natural resource, which is available in all human societies – even in the poorest ones.

Commercial fertilizers are unaffordable in many poor societies, and urine may be one way of increasing food production with the help of a cheap and readily available nutrient source.

"If human urine with good microbiological quality could be utilized for plant production, millions of people living in the tropics or semi-tropics including the poor or the poorest of the poor, as they are called in Bangladesh, could increase yields of edible and non-edible plants cultivated in small plots or even in pots," the researchers say.

They point out that anyone attempting to exploit the benefit of urine as a fertilizer needs to be aware that the nutrient quality and quantity vary. They found that the chemical composition of urine depends on time of day, diet, climate, physical activity and body size.

They caution too that urine from men and women in tropical developing countries where the amount of drinking water is limited and sweating is high may be more concentrated. Increased sweating means that more nitrogen is lost through the skin so less gets into the bladder.

Dr Cynthia Mitchell of Australia's Institute for Sustainable Futures says that urine will soon be too precious to flush down the loo: "It is time to put our 'flush it out of sight' mentality aside and see the very real social, environmental and economic benefits of recycling poo and wee," she says.

She points out that the world is fast running out of mined phosphorus for fertilizers, and urine may be the answer. Cities are becoming phosphorus hot spots because of urine in sewage, she says, while global underground reserves of phosphorus are unlikely to last more than 50 to 100 years.

Not to be outdone by the urine researchers, academics at the Technical University of Denmark have been looking at the composition of human excreta in southern Thailand and its use as a fertilizer. They have calculated the frequency of defecation for people on different diets – 2.4 times a day for those having a curry lunch compared with once a day for pork soup.

Average daily output was 120 to 400 grams a day, and it was found to be full of nutrients: "Of the total amount of elements in waste human excreta accounts for the following fractions: 75% of the nitrogen; 50% of the copper; 40% of the phosphorus, potassium and sulphur; 20% of the calcium, magnesium, lead; and less than 10% of the zinc, nickel, cadmium and mercury," say the researchers.

They add, "The potential of human excreta as a fertilizer should be utilized as a valuable resource, rather than a waste product."

The only downside to using excreta is that it contains toxins as well as nutrients. The presence of toxins also means that if urine is to be used, it would need to be kept separate from excreta. A number of centres in Sweden have been experimenting with equipment that does separate the two.

Smelling body odour

UNDERARM BODY ODOUR CAN BE a way to a man's heart.

At certain times of the month the female armpit releases a different, more pleasant odour designed to appeal to the passing male nostril, according to researchers who spend much of their time analysing body odour.

With the help of samples of underarm body odour and a panel of male volunteers, academics have been able to show for the first time that at the fertile time of the monthly cycle, a woman's body odour changes and becomes more attractive.

They say the results suggest that body odour may be a way in which women, unknowingly, signal their fertility, with changing hormones also contributing to the changing odour.

In the research, scientists looked at axillary or armpit odour changes across the menstrual cycle, to see whether or not women do have fertility signals.

"Females of a number of primate species display their fertile period by behavioural or physical changes. Traditionally, the fertile period in human females has been considered to be concealed, but this presumption has rarely been tested," says Jan Havlıcek, an anthropologist at Charles University, Prague, who led the study. "Our results suggest that men can potentially use axillary smell as a mechanism for monitoring menstrual cycle phase in current or prospective sexual partners."

In the research, 12 women wore cotton pads in their armpits for 24 hours at a time. These were collected, and kept in sealed jars ready for 42 men aged 19 to 34 to act as testers. None of the women was taking the pill during the study, and they were also asked to refrain from using perfumes, deodorants, antiperspirants, aftershave and shower gels, and not to eat meals containing garlic, onion, chilli, pepper, vinegar, blue cheese, cabbage, radish, fermented milk products or marinated fish. Sexual activity was also out, as was sleeping in the bed of a partner while wearing the pads.

The results of the men's rating show links between attractiveness and fertility, and times of the month. The odour of women in the follicular phase was rated as the least intense, the most pleasant and the most attractive.

"Our results show that both the pleasantness and attractiveness ratings given were lowest during menstruation and peaked in the follicular phase when the probability of conception is highest," say the researchers.

The opposite pattern was found for odour intensity. It was most intense during the menstrual phase and the least intense during the follicular phase.

"We demonstrated for the first time that armpit odour itself may carry information about women's fertility status," say the researchers.

Other research shows that gay men and women had different preferences from heterosexuals of both genders, while researchers at the University of Liverpool have found that identical twins may share body odour as well as almost everything else.

Their research shows that there is also a genetic influence on body odour. The researchers say body odour is thought to be used during mate choice to select genetically compatible mates.

To test the idea, they investigated odour in twins: "We show that odours of identical twins can be matched by human sniffers at rates better than chance, even when the twins are living apart. These results indicate an important genetic influence on body odour."

But body odour may not be the only signal of fertility. One study has found that facial images of women taken during their follicular phase were seen as more attractive compared with images of the same women taken during the luteal phase.

Other research also suggests that ovulation is not as hidden as previously assumed. It has been found that women become more symmetrical at that time of the month; their waist–hip ratio is slightly lower and skin becomes lighter.

Another study, based in a discotheque rather than the laboratory, found that women dress differently too. Married or partnered women who were unaccompanied at the disco were observed by academics to be dressed in a more sexually provocative way when they were in the most fertile period.

Shirt smellers

SMELLING MEN AFTER THEY HAVE eaten meat is all in the line of duty for anthropological researchers.

Men who eat meat smell differently to other men, it appears, and it's not very attractive.

Women volunteers who sniffed the body odour of men devouring red meat, and that of non-flesh-eaters, found that the odours of the meatless men were more pleasant, more attractive and less intense.

The underarm odour from the meat munchers was judged to be stronger and more intense.

"The results of this study show for the first time that red meat consumption may have a perceivable impact on body odour," say the anthropologists from Charles University in the Czech Republic. "We suggest that the main non-genetic source of underarm variation in a healthy human is due to differences in diet."

It's known that body odour is very individual and that it can be affected by a variety of factors, from medication and disease to fear and hormonal levels.

In the study, men volunteers consumed meat or non-meat diets for two weeks at a time. Other than meat content, the meals were identical, and the meat eaters had 200 g a day spread over two meals.

The men used unperfumed soap, and were asked to avoid physical activities, sexual activity and sleeping in the same bed as their partner. At the end of the two-week period, the men wore cotton pads under their arms for 24 hours. The pads were then recovered and given to 30 women, who were asked to rate them on a seven-point scale in a well-ventilated room.

"Results of repeated measures showed that the odour of donors when on the non-meat diet was judged as significantly more attractive, more pleasant, and less intense. This suggests that red meat consumption has a negative impact on perceived body odour hedonicity," say the researchers.

They say it's not possible to say how long the meat content in food remains perceptible in body odour, or how much meat is needed to have an effect.

It's also not known how eating meat could have such an effect:

"Current knowledge allows us only to speculate what particular compounds and metabolic processes are responsible for hedonic changes in body odour after the meat consumption," they say.

One theory is that the way the body processes fat from meat may be implicated in some way.

But genes could be involved in body odour too, according to a study at the Twin Research Unit at Guy's and St Thomas' Hospitals and Newcastle University. It found that genes may play a big part in body odour, which would make each individual's odour unique.

The study, which involved human "sniffers" attempting to match pairs of twins from sets of five different body odours, found that the human nose could not differentiate between separate samples from identical twins and two samples from the same person. Identical twins who shared all their genes were rated as much more similar than pairs of non-identical twins who shared only 50 per cent of their genes like brothers or sisters.

The findings, published in *Chemical Senses* journal, raise the possibility that body odour could be used as a sort of olfactory fingerprint in forensic applications.

Professor Tim Spector, director of the Twin Research Unit at St Thomas' Hospital, says: "This study adds to the increasing evidence that genetics play an important factor in body odour and in the future discovery of the genes involved has the potential to be used to aid forensic investigations and help with non-invasive diagnostic tests – and may one day possibly make people smell nicer!"

Toenail clippings

YOU JUST NEVER KNOW WHEN those old toenail clippings could be useful, all 100,000 of them.

Only the most enthusiastic of researchers could have foreseen the eventual importance of cataloguing and storing individual clippings taken from more than 60,000 women half a century ago.

Just how long it took to take the clippings, and how they were catalogued, is not clear, but what is now known, thanks to them, is that toenails can reveal the long-term risk of heart disease.

Researchers at Harvard University analysed all nail clippings taken from 62,000 women, and then compared the findings for the 900 women who had gone on to develop heart problems with those for the women who had remained healthy over the 25-year period.

Results of the research show a link between nicotine levels in the nails and risk. Women with the highest levels of nicotine were 3.4 times more likely to go on to have heart disease. For each unit increase in nicotine content, the risk rose by 42 per cent. "Toenail nicotine levels are predictive of heart disease among women independent of other risk factors and remained significant even after adjusting for history of cigarette smoking," say the researchers.

The Harvard team are not the only researchers to clip and keep toenails. In a second study, based on nearly a thousand men from ten European countries, including the UK, researchers found that levels of a natural compound called cerium were linked to risk of attack. Results show that those with the highest levels of cerium in their toenails were twice as likely to have a heart attack.

"Our results suggest that toenail cerium levels may be associated with an increased risk but more research is warranted to fully understand the plausibility and public health implications of these findings," say the researchers.

And heart disease is not the only health problem that toenail clippings analysis can be used to detect or predict. Levels of selenium, a trace element that works as an antioxidant, can also be used to detect risk of prostate cancer.

Researchers at Maastricht University, who looked at the toenail clippings of 58,000 men aged 55 to 69, who were then monitored for more than six years, found that those who had had the highest toenail level of selenium were less likely to develop the disease – a third less likely, in fact. "Results confirm that higher selenium intake may reduce prostate cancer risk," say the researchers.

Selenium levels in toenails may also be a marker for risk of pre-eclampsia, a condition that affects women in pregnancy and whose main feature is high blood pressure. In research at the John Radcliffe Hospital, Oxford, 53 patients with the condition and 53 healthy pregnant women donated toenail clippings months before giving birth. Results show that toenail selenium levels in the patients were significantly lower than those of the healthy women. Women with the lowest toenail levels were 4.4 times more likely to have the condition.

Stomach cancer risk too can be seen in toenails, according to a study at Kagoshima University in Japan. It shows that the higher the zinc levels in toenail clippings, the lower the risk, especially in smokers.

Toenail analysis can be used to detect occupational exposure diseases – more than 25 compounds were found in the nails of Italian construction workers in one study. It can also be used to detect drug abuse.

Smelling fear

SCIENTISTS HAVE DISCOVERED THE SMELL of fear, and it is infectious.

With the help of scary movies and samples of body odour, they have shown that the smell of someone else's fear can affect the behaviour of others, making them more cautious, accurate and alert.

"This is the first study of the effect of human fear chemical signals on performance. There has not been any evidence that human chemo signals release immediate behaviour," say the researchers, whose findings are reported in the *Chemical Senses* journal.

While animals, from anemones, earthworms and minnows to mice, rats and deer, are known to communicate fear through chemicals in body odour, it's the first evidence of such an effect on behaviour in humans.

Research shows that experience of fear is accompanied by a series of neurochemical changes, some of which may be found in the sweat. These changes are then thought to have an effect on behaviour.

In animals, chemicals are released when they are under stress, and can act as warning signals to generate an increase in vigilance in animals of the same species, and can also trigger changes in the immune system.

In the research, men and women volunteers were given a questionnaire detailing 50 popular films of varying emotional content. Each then answered questions about how the video made them feel, and they then watched 20-minute segments of movies rated scary or non-scary.

At the end of each segment, they described how they felt, whether they were sad, angry, anxious, afraid, disgusted or neutral. Heart rate and other data was also collected.

The volunteers, who were videotaped with a hidden camera during the experiments, also wore pads under their arms to absorb any chemicals produced by the body. At the end of the research, the pads were removed and stored.

In the second part of the experiment, these pads were taped between the nostrils and mouths of 50 other volunteers, who were first asked to rate the intensity and pleasantness of the smell, and then given a word association task.

How they performed that word task was measured and recorded

and compared with the type of pad that the volunteers had been exposed to. The pads were put into three different classes – those taken from people who had been fearful after watching the movies, those taken from people who had been content, and a no-sweat or control.

The results showed that those who had been smelling the fear of another person were more cautious and more accurate when they completed the word tests compared with the others.

Possible factors that could introduce bias, including individual differences due to anxiety, verbal skills and the qualities of the smells, were ruled out by the researchers from Rice University in America.

The researchers say that the effect may serve to prime people to be on the alert: "We have tested the effects of fear chemical signals on cognition and attention. We showed that fear heightened caution and vigilance. Overall, we found that participants in the fear chemical signal condition tended to be slower and more accurate," say the researchers.

"Our results demonstrate that human fear chemosignals enhance cognitive performances in the recipient. Those in the fear condition behaved as if they were motivated to avoid misses. Such learned association in humans may prime people to be on the alert."

COUNTING KISSES, FAST FOOD SHOPS, STAIRS, YAWNS AND HOW MANY TIMES PEOPLE SAY SORRY

Fast food shops

COUNTING FAST FOOD SHOPS MAY seem an unusual occupation, but it could have important implications for the health of the nation.

For researchers have discovered that deprived areas have five times as many fast food shops as affluent parts of the country. One in three fast food shops are located in the poorest areas, which also have the highest rates of obesity and other diet-related health problems.

According to the research, there are now more than 2,500 fast food chain shops in England and Scotland belonging to the big four – McDonald's, Burger King, Kentucky Fried Chicken (KFC) and Pizza Hut.

The study, by the Medical Research Council's Social & Public Health Sciences Unit in Glasgow, and Queen Mary College, University of London, shows that the number of outlets increases in proportion to the levels of deprivation.

"People living in deprived areas are doubly disadvantaged. They have to contend with low income and with reduced opportunities for healthy eating," the researchers say. "Results provide support for a concentration effect where health-damaging risk factors for obesity appear to be concentrated in more deprived areas of England and Scotland. It may partly explain geographical variations in overweight and obesity, and why rates are higher in deprived areas."

Obesity rates are not only high but rising, leading to a predicted increase in obesity-related diseases, including type II diabetes, heart disease and high blood pressure.

Research has shown a link between obesity and overweight and neighbourhood deprivation, even when social class, income, age and gender are taken into account.

"It has been suggested that residents of deprived neighbourhoods have greater exposure to certain factors which may facilitate the development of overweight. Consumption of fast food, in particular, has been associated with increasing overweight and it has been hypothesized that fast food and other outlets which sell energy dense, high fat foods at a low price might be more prevalent in deprived areas and that this might partly explain the greater prevalence of obesity in these areas," say the researchers.

In the study, the researchers looked at the locations of each of the branches of the four brands, the levels of deprivation in those areas, and the populations. They then calculated the ratio of branches to population.

There were five levels of area deprivation or affluence, and the results show that for each of the brands there was a much greater concentration of restaurants in the fifth, most deprived group of areas. There are 845 outlets in the most deprived group, compared with 188 in the most affluent.

When the researchers worked out the number of restaurants for each 1,000 people in the population, there were around five times as many restaurants in the most deprived areas compared with the most affluent.

According to the report, KFC had the smallest presence in the most affluent areas of England, and the largest in the poorest. Deprived areas have ten times the number of KFC restaurants as affluent areas for the same population. For McDonald's and for Pizza Hut, there are fourfold increases in deprived areas, and for Burger King the difference is threefold.

"Our results add to the limited literature which links higher densities of chain fast-food restaurants to more deprived neighbourhoods and hypothesizes that such outlets may act as a discriminator in terms of quality of the local food environment," say the researchers.

Just why there is such a difference is unclear: "One of our main observations is that there is a linear increase in response to deprivation. In other words, the more deprived an area gets, the more of these kinds of stores it gets," says Steven Cummins.

"A number of factors may be involved in this. It may be because deprived areas are commercially desirable because, for example, there might be a greater local demand for fast-food, or it may be because land prices might be cheaper."

Stair counters

USING THE STAIRS RATHER THAN the lift can significantly lower levels of bad cholesterol, increase oxygen consumption, and lower heart disease risk. Young women who took part in a stair-climbing exercise saw big health benefits within eight weeks without any change in their diets or lifestyle.

The women, described as sedentary but otherwise healthy at the start of the exercises, also increased their oxygen consumption by almost 20 per cent.

In the research at the University of Ulster and Queen's University, the women were put through an exercise regime of short bouts of stair-climbing to assess its effects on heart fitness and cholesterol levels.

The stair-climbing started with one ascent a day, for five days a week, of a public staircase with 199 steps. The women were set the task of climbing the steps at a rate of 90 steps a minute, taking around two minutes a time.

Eventually the women were climbing the steps five times a day, five days a week. They all agreed at the start of the research not to change their diet or lifestyle over the experimental period.

The results show that, compared with a group of similar young women who did not do the exercises, those who did had a 17.1 per cent increase in oxygen consumption and a 7.7 per cent decrease in bad cholesterol levels. There was no change in total cholesterol levels.

Maximal oxygen consumption is important because it represents the amount that can be used by the body for maximum power. The more oxygen you can consume, the more energy, power or speed you can produce.

"The study confirms that accumulating short bouts of stair climbing activity throughout the day can favourably alter important cardiovascular risk factors in previously sedentary young women. Such exercise may be easily incorporated into the working day and therefore should be promoted by public health guidelines," say the researchers.

According to researchers at the University of Edinburgh, regular stair-climbing not only provides free daily exercise and burns more calories per minute than jogging, but it has been shown to provide other important health benefits. "Research shows that just seven minutes of stair climbing a day cuts the chances of dying from coronary heart disease by 62 per cent and halves the risk of a heart attack," they say.

Counting kisses

TURNING THE OTHER CHEEK WHILE kissing may reveal more than you think.

Academics who have been spending their time watching people kiss have discovered that eight out of ten people, whether they are left or right handed, turn their heads to the right when they kiss.

After observing couples in public places, and also inviting volunteers to kiss large dolls, they found that four times as many men and women homed in on the left cheek as the right.

Just why is not clear, but there are two theories. The first is that humans are hard wired to favour movements to the right almost from the moment of conception, while the second is that moving to the right exposes the left side of the face, which has been linked to the right, emotional side of the brain.

"One theory that has been put forward is that by turning their head to the right, the individual reveals their left cheek which is controlled by the emotive right cerebral hemisphere," says Dr Julian Greenwood, head of science at Stranmillis University College, Belfast, who led the study.

"Even though we may be aware of these various kissing cultures and behaviours, we don't stop to think about how we kiss: will I kiss on the lips or on the cheek? or, will I turn my head to the right or the left? Instead, kissing is a spontaneous act, although one that is dependent on context: friends may kiss on the cheek while spouses and partners may kiss on the lips."

The research involved scientists making observations of kissing couples: a single kiss on the cheek or the lips was recorded where there was a clear head turn to the right or to the left. In the case of several kisses, only the first was recorded.

In the second part of the work, 240 volunteers were asked to stand directly in front of a large (about two-thirds life-size) symmetrical doll's face and kiss the doll on the cheek or lips. Results showed that 80 per cent of men and women turned to the right.

"The results of the study showed a clear predominance for people to turn their heads to the right when kissing, whether kissing another person or kissing a doll's face, with no significant difference between the two situations," says Dr Greenwood.

Just why is not clear, and, as mentioned, there are two competing schools of thought. The motor theory is that humans naturally turn to the right, while the emotional theory is that it is all about exposing the left, more emotional side to the recipient.

"It is clear that a motor bias causes a turn to the right in the majority of individuals: but is this turn modified by the emotion of the kissing situation?" says Dr Greenwood.

He says that one clue is the high rate of right turning when kissing dolls: "As there was no difference between kissing couples and doll-kissers it might be assumed that the motor theory provides a better explanation for kissing behaviour than the emotion theory."

But in another study involving head-turning in portraits, researchers who asked men and women to pose for either an emotional family portrait or an impassive portrait that avoided depicting any emotion at all found that the sitters were more likely to turn their left cheek in the emotive condition and their right cheek in the impassive condition.

Professor Lesley Rogers of the University of New England in Australia, and an expert on asymmetries in brain function and behaviour, says, "When they are turning their faces to the right, people line up the left-hand side of their faces which is controlled by the right hemisphere of the brain which governs intense emotions."

Researchers at the University at Albany have also been counting kisses and found that out of 1,041 college students, only five had never experienced romantic kissing and more than two hundred had kissed more than twenty partners.

According to the research, kissing between sexual or romantic partners occurs in more than 90 per cent of human cultures, as well as in chimpanzees.

"Although kissing is a widespread practice among humans, few scientists have attempted to assess the adaptive significance of kissing behaviour," say the researchers, whose study shows that the information conveyed by a kiss can have profound consequences for romantic relationships. It shows that many college students find that after they have kissed for the first time they are no longer interested.

"In other words," say the researchers, "while many forces lead two people to connect romantically, the kiss, particularly the first kiss, can be a deal breaker. Kissing is part of an evolved courtship ritual. When two people kiss there is a rich and complicated exchange of information involving chemical, tactile, and postural cues. This may activate evolved mechanisms that function to discourage reproduction among individuals who are genetically incompatible."

The study also found sex differences in the importance and type of kissing. Men tended to kiss as a means to an end – to gain sexual favours or to reconcile – while women kiss to establish and monitor the status of their relationship, and to assess the level of commitment on the part of a partner.

Women were found to be more likely than men to insist on kissing before a sexual encounter, while men were more likely to initiate open-mouth kissing and kissing with tongue contact. The researchers suggest one reason for that is that male saliva contains the sex hormone testosterone, which can affect libido.

Counting offsides

RESEARCHERS HAVE FINALLY DISCOVERED WHAT football fans have long suspected ... linesmen get it wrong too often.

Almost one in four offside decisions are wrong, according to academics, and in the first fifteen minutes of a game almost 40 per cent of decisions are not right.

The errors may, it's suggested, be down to optical illusions or to the flash-lag effect, wherein a fast-moving object appears to be up to a metre ahead of its real position.

"The incidence of errors in applying the offside law suggests that alternative ways for improving offside assessment should be considered," say the academics. "As a result of the increased financial reward, along with the worldwide broadcast of football, the decisions made by the match officials during matches very often come under close scrutiny, especially if an incorrect decision has a direct impact on the result."

Until now, however, very few attempts have been made by researchers to examine the correctness of match officials' decisions during matches, and the factors that may impact on these decisions.

In the research, part funded by FIFA, the academics analysed all the games of the 2002 World Cup. Videos were painstakingly converted into digital images and offside decisions and the positions of players and linesmen or assistant linesmen recorded and analysed.

The results of the research show that across all the matches played, there were 337 offside decisions, an average of just over five per game. The linesmen got it wrong 26.2 per cent of the time. Most errors involved wrongly raising the flag, rather than failing to flag a genuine offside. Almost nine out of ten of the errors involved wrongly flagging a pass that was not offside.

The researchers say the error rate was high, and point out that in just that one tournament errors led to the cancelling of five legal goals, the awarding of two illegal goals, and ruining four goal-scoring opportunities.

The researchers from Belgium's Katholieke Universiteit Leuven found that the error rate was highest during the first fifteen-minute period of the match, at 38.5 per cent. The second highest was 29 per cent in the first fifteen minutes of the second half.

"Assistant referees appear to need some time to perceptually get used to the typical movements of the defenders and attackers around the offside line," says the report.

All the evidence, they say, suggests that linesmen are being affected by what they call the flash-lag effect. Defenders make the effect worse for the linesmen when they run in the opposite direction to the attacker. That makes the linesman see the attackers as being even farther forward than they really are.

While proving fans right, the researchers have shown another terrace theory to be wide of the mark. Linesmen are not influenced by players raising their arms to claim offside, they found. The results show that the number of players claiming offside was the same for correct and incorrect decisions.

Counting chores

BRITISH COUPLES SPEND LITTLE MORE than two hours a day on housework.

Research based on men and women in 34 countries shows that couples in the UK spend 19 hours a week on housework.

That's the fifth-lowest rate, and it's dwarfed by the 47 hours put in by couples in Chile, who top the league table, those in Ireland (40 hours) and Germany (30 hours).

According to the report, women in Britain still do 71 per cent of the housework, with men doing less than a hour a day, half the time put in by Russian males.

The research, reported in the *European Sociological Review*, found that the number of hours put in by women varies with levels of empowerment, while men's changes with wealth.

Women's hours go down when they have obligations outside the home, while men's go down with increasing wealth.

"We argue that women's housework efforts might be more sensitive to female empowerment, and men's to the dynamics of economic activities," say the researchers from the University of Stavanger and the Institute of Sociology, Bergen, Norway.

The housework survey, based on more than 17,000 people in all 34 countries, asked how many hours they and their spouses put in each week, not including childcare.

Results show that couples in France put in the fewest hours – 15.9 – followed by Norway (16), Finland (17.8) and Britain (19.4). Chile topped the league with 47.6 hours, followed by Brazil (43.6), Mexico (43.1), Ireland (40.3) and Russia (38.4).

The results also show that among women, Chileans put in the most hours – 38 – and Norwegians the fewest, 11.67. Russian men put in the most housework hours – 13 – and Japan the fewest, 2.5, or around 20 minutes a day.

"Husbands in Japan do hardly any housework compared with their wives, while men in the Philippines, Russia, and Mexico do relatively more," says the report.

This lack of effort on the part of Japanese men makes Japan the country where women do the biggest share of the housework – 91 per cent. The most egalitarian countries were Latvia, Poland and the

Philippines, where men did the biggest share of housework – 36 per cent.

"Wives who have higher income than their husbands or have less time available from work outside the home share housework less unequally with their spouses. An egalitarian gender ideology is also associated with women doing relatively less at home," adds the report.

How many times people say sorry

SAYING SORRY IS A MIDDLE-CLASS THING. The middle classes apologize twice as much as the working classes, and men over 25 say sorry more often than anyone else.

According to the research, the powerful also apologize more, and that may be a way of looking good, and identifying with inferiors, but at the same time underlining who is really in charge.

According to the research, people in Britain apologize on average five times a day, but for some people, as many as one in every 150 words spoken in a typical day may be an apology of some kind.

Researchers say the results show that apologizing in Britain is linked to class: "The use or avoidance appears to be an important way of signalling class allegiance. We found that working-class speakers do apologize, but given similar circumstances, they do so to a lesser extent than middle-class speakers. Apologizing in Britain may be a way of signalling your social identity," says Dr Mats Deutschmann, who led the study.

In the research, taped recordings of everyday conversations from a collection of 4,700 men, women and children in Britain were used, with a more detailed analysis of 1,700 of the speakers. The recordings covered a minimum of two days' total conversation at home and work, and key words looked for included "afraid", "apologize", "apology", "excuse", "forgive", "pardon", "regret" and "sorry". Dialogue of speakers of known age, gender and social class was then analysed for any differences.

The results show big class and age as well as other differences in the rate of apologizing. Overall, the middle classes apologized 93 times in every 100,000 words spoken, compared with 42 times among working classes.

Middle-class men and those aged 34 to 44 apologized the most – 125 apologies for each 100,00 words – while working-class men over 45 apologized the least – 32 times in 100,000 words.

Middle-class men apologized more than women. Men aged over 45 had 76 apology words per 100,000, compared with 45 for middle-class women of the same age. In the working classes, women apologized slightly more than men. The results also show that, overall, apology use declines with age.

The results also show that women increased their apology frequency in formal situations, whereas men were more likely to say sorry – the most used apology word – in informal situations.

The over-45s were more likely to offer casual apologies, while "sincere" apologies were more favoured by middle-class speakers. Women also apologized more for "accidents" than did men, and the researchers say that this is mostly a result of gender differences in the tasks performed at home. Men on the other hand apologized more for social gaffes.

"The analysis indicated the existence of social-class differences in politeness norms in Britain. Middle-class speakers tended to apologize more for social etiquette breaches – 'Lack of consideration' offences – and were more inclined to apologize for disagreements arising from misunderstandings," says Dr Deutschmann of Mid Sweden University.

"Working-class speakers, on the other hand, apologized more for 'Social gaffes', especially for more serious breaches such as belching.

"Low-status groups – females, and young speakers – were more inclined to use strategies which involved the acknowledgement of responsibility for an offence. High-status groups (males and older speakers), on the other hand, showed a greater tendency to use strategies that minimized responsibility."

Just why middle-class men and men in powerful positions apologize more is not clear, but there are a number of theories.

"The apology [represents] important ways for the speaker to preserve a positive image in the eyes of the less powerful. While acts of politeness have traditionally been ways for social inferiors to show deference towards social superiors, the latter group have come to use them towards social inferiors as a sign of solidarity or favour.

"Arguably, the unexpected distributions of apologies, whereby the powerful used the form more often than the powerless[, demonstrates] important ways for the speaker to preserve a positive image in the eyes of the less powerful.

"In reality, however, this egalitarianism is an illusion. In multinational companies, on the political scene, and in the public sector it is the privileged classes who are still in charge. One strategy for minimizing the gap between pseudo ideals and the real state of affairs is for the powerful to appear 'humble' when confronted with the less powerful; downward politeness is one expression of such a strategy. As this mode of rhetoric becomes the norm, downward politeness paradoxically becomes a linguistic marker of power, and a tool for exercising that power.

"Using polite formulae in public situations is arguably a way of presenting oneself as 'respectable', not only to the addressee, but also in the eyes of others present.

"The results show that social-class differences in apologetic behaviour do exist in Britain. The use or avoidance of polite forms appears to be an important way of signalling class allegiance."

The researchers add, "A very rough calculation seems to suggest roughly 4–5 apologies per person and day. A more accurate indication is that the Brits apologize roughly 70 times per 100,000 words spoken. An average utterance, in informal conversations, is around 10 words, which basically means that 7 utterances of 1,000 are apologies."

Counting yawns on trains

IN THE WORLD OF YAWNING RESEARCH, it has been one of the big unanswered questions: Do men yawn more than women?

In all the species of non-human primates for which data on the frequency of yawning has been reported, males yawn much more than females, and one theory is that it's a sign of dominance and potential aggression.

This theory is based on the fact that in all these species, the males have bigger canine teeth. Since the most serious injuries in primates are usually caused by biting, yawning may therefore be an intimidating display of the teeth toward potential antagonists. But is the same true in humans, where the teeth of men and women differ only slightly?

We start yawning in the womb as early as eleven weeks after conception; the average yawn lasts six seconds; we yawn up to 3.4 times a minute when we are bored; and yawning is highly contagious. But much about yawning, including its evolutionary purpose, remains unknown.

In research at the Università di Roma, Italy, academics set out to test the hypothesis that sex differences in yawning are related to size differences in canine teeth. A further aim of the study was to get the first data on how often people yawn in natural situations.

"A detailed description of human spontaneous behaviour seems to be a necessary first step to reach a full comprehension of the cause and function of human behaviour," say the researchers.

In the study, the researchers counted the yawns of passengers of the B line of the Rome underground. One of the reasons they chose that line was that the trains have no usable windows, so passengers looking at each other is considered normal behaviour. Trains were also chosen because they are environments in which passengers do not know each other, which means they were more likely to show assertive behaviours, including yawning, if it is a sign of assertiveness.

Data was collected during 94 underground journeys, each lasting around 15 minutes, over a 12-month period. At the beginning and end of each journey, the observers counted the men and the women present in the carriages, as well as the yawns.

For the purposes of the study, the observers were given a concise definition of a yawn: "Yawning is a gaping movement of the mouth accompanied by a long inspiration followed by a shorter expiration."

The observers were also asked to distinguish between the two types of yawn – one where there is no attempt to hide or cover the yawn, and the other where the man or woman partially or completely covers the mouth with a hand, or yawns incompletely with a partially closed mouth.

A total of 267 yawns by 221 different people was recorded; and results show there was no significant difference between the sexes: "Data indicate that adult human males and females, unlike several non-human primate species, do not differ in the frequency of yawning," say the researchers.

The results do show, however, that uncovered yawns were more frequent in men, lending some support to the theory of yawning as an assertive action. A total of 49.2 per cent of men and 32.6 per cent of women who yawned showed uncovered yawns, suggesting that yawning has evolved as more of a display among men than women.

"The most conspicuous form of yawning is, thus, used more by males than by females. Hence, human males seem still to behave in a more assertive way, even if they do not appear to differ from females either in canine size or in the total frequency of yawning," say the researchers.

There could of course be another explanation, as the researchers readily admit: "Alternatively, human females may be more polite than males."

Totting up high heels

WEARING HIGH HEELS MAY BE healthy after all.

Doctors have discovered that women who wear heels of up to seven centimetres may have healthier pelvic muscles and be less likely to get common pelvis and low back pain.

Research based on 66 women shows that heels increase the angle of the foot, with a beneficial effect on the muscles.

"Wearing high heels results in pelvic floor muscle relaxation, thus improving pain symptoms," say urologists at the University of Verona, Italy.

High heels have long been associated with health problems, including corns and calluses, hammer toe, bunions, toenail problems, stress fractures, joint pain, shortening of the Achilles tendon, heel inflammation, and abnormal growth of nerve tissue.

But now the Italian researchers have found what is thought to be the first health benefit of heeled shoes.

In the research, reported in the medical journal *European Urology*, doctors measured the electrical activity in the pelvic muscles of the women, all aged under fifty, when they held their foot at different angles,

An electromyographic (EMG) biofeedback instrument with electrodes attached to the skin surface was used to monitor changes in pelvic floor muscle activity at 5, 10 and 15 degrees. Results show that the electrical activity was some 20 per cent lower at 15 degrees.

That 15 degrees is equivalent to a heel size of 7 centimetres.

The researchers believe that raising the ankle on a heeled shoe results in a pelvic tilt that boosts the contracting power of the pelvic muscles, reducing pelvic pain.

"Two years ago we planned and started our study in order to corroborate the hypothesis that variations in ankle inclination might affect pelvic floor muscle performance," they say. "The preliminary study results would seem to corroborate our initial hypothesis.

"A relaxation of the pelvic muscles induced by heels might have beneficial effects, reducing the burden of chronic pelvic pain. Our preliminary study results lead us to suppose the strain on ankles induced by heels – three to seven centimetres – might facilitate pelvic floor muscle performance. A standing posture with heels would result in a posterior pelvic tilt able to reduce pelvic floor muscle tone, thus producing a relaxation of the muscles."

WATCHING ANTS, COUNTING WORMS, LISTENING TO DOGS, AND GRUNTING AT BABOONS

Watching ants find their way home

ANTS SEARCH FOR FOOD OVER relatively vast distances.

Leafcutter ants, for example, travel up to a quarter of a mile from home, and they often make the journeys at night or when it's overcast.

Night-time foraging means that use of celestial cues and landmarks is very difficult, and overcast days would also rule out a role for the sun. So how do they find their way back to the nest?

Many insects use different techniques to orient themselves. Some use so-called geocentric cues like landmarks, odours and other features from which they are able to determine their current position relative to that of the nest. Celestial cues, including the sun's azimuth and polarized skylight, are used too.

None of these seemed likely for the leafcutter ants, and the researchers at the Smithsonian Tropical Research Institute, Panama, suspected that the invertebrates might have some kind of on-board natural magnetic compass that would show them the way home.

To find out, they tracked a colony of ants with a cunning plan to try to make the insects go home the wrong way.

"Leafcutter ants are active when and where skylight and the sun are not visible. They forage in the deep shade of the forest in overcast conditions or during night. For this reason, we investigated whether leafcutter ants use a magnetic compass to orient and path-integrate," they say.

In their studies in the tropical rainforests, the researchers track down large nests and map the routes the ants take. They then use bait to divert foraging trails by scattering oat flakes in a line to a large four-coil electromagnet covered with black vinyl and housed in a 6-metre-long plastic tent, along with a table covered with sand.

The ants were intercepted on their foraging route as they headed north for home with food. They were taken south and put inside the completely darkened electromagnet tent.

A control group of ants was tested in the natural geomagnetic field

– which in Panama is oriented towards geographical north – and the experimental group was exposed to a magnetic field of reversed polarity. The movements of the ants were then tracked with an infrared camcorder to see which way they went.

In a second type of experiment, the team disrupted the ants' magnetic compass with a strong magnetic pulse to trigger a change in their ability to find their way home. In this study, ants were intercepted and carried to an electric coil through which a weak current passed to produce a weak magnetic field.

Both studies show that the ants use a sophisticated magnetic compass to navigate. In the first experiment, all the control ants went home as usual, but half the second group went the wrong way.

"We have shown that they can rely on magnetic information to orient their path home vector. In our first experiment, night-foraging workers responded to a local reversal in the polarity of the geomagnetic field," say the researchers.

"Our results show the use of a magnetic compass during path-integrated navigation in leafcutter ants and provide insights on the nature of the particles forming the compass. This is the first evidence of the magnetic compass being involved in path integration in an invertebrate."

Measuring the genitalia of the polar bear

CLIMATE CHANGE AND GLOBAL POLLUTION are major environmental problems threatening the very future of life on earth.

But how do you measure it? What could be used as a barometer for how badly or how well things are going? Is there anything out there that could give a regular measure?

One answer, it seems, could the genitalia of polar bears, whose size may be influenced by global pollution and climate change.

A team of scientists has discovered that the genitalia of bears that are exposed to the greatest levels of pollution are significantly smaller than those living in areas where pollution is the lowest.

They warn that the combined effects of pollution and climate change threaten to damage the reproductive capability of the bears.

"The worst case outcome is that the combined effects from shrinking ice coverage and pollution on reproductive success could be extinction of the polar bear," says Dr Christian Sonne, Senior Research Scientist at the National Environmental Research Institute and Department of Arctic Environment in Denmark.

A large baculum – penis bone – is, of course, extremely important when mating in an arctic climate, so even a relatively small reduction in length may have an impact on future polar bear populations. Such a reduction in penis size could make copulation less successful and, in a worst-case scenario, result in a population decline because of the already low reproductive rates of the species.

The lifespan and adult survival rates of polar bears are among the highest for mammals in general, while the reproductive rates are among the lowest for terrestrial mammals. Polar bears are seasonal breeders, exhibit delayed implantation, and usually give birth to two cubs. Mating takes place in April to June, and implantation occurs in late September, resulting in birth in late December.

In the research, Dr Sonne and colleagues looked at the sizes of the genitalia of male and female bears in East Greenland and compared them to those from two other areas, Svalbard and Canada. Testicles and penis, and ovaries and uterus, were all examined, weighed, measured and recorded.

Results show that the East Greenland bears' genitalia were smaller and lighter than those of the bears from Svalbard, which were in turn smaller than those from Canada.

The researchers also found that the size was linked to levels of pollution of endocrine disruptor chemicals such as PCBs. The human endocrine system is a network of glands and hormones that regulates many of the body's functions, including that of the testes and ovaries. An endocrine disruptor is a synthetic chemical that mimics or blocks hormones, disrupting normal working.

All the Svalbard polar bears that were included in the present study were not exposed to pollution, while the East Greenland bears were heavily polluted with a toxic cocktail of organochlorines, polybrominated flame retardants, mercury and perfluorooctane. Polar bears from the Canadian Arctic were significantly less affected by pollutants than were the East Greenland bears.

The researchers say one theory is that long-range transported endocrine-disrupting pollutants such as PCBs, DDT and mercury could have a negative impact on the size of sexual organs in polar bears.

They say that polar bears from East Greenland are polluted because they rely on blubber from mainly ringed seals (*Phoca hispida*) and bearded seals (*Erignathus barbatus*). The seal blubber contains significant amounts of toxic compounds which may have an effect on hormones and vitamin levels in the bears.

"The important question is whether the impact from future climate changes – food accessibility – and long range transported persistent organic pollutants is a threat to polar bear reproductivity and thereby the species," says Dr Sonne.

"Our work suggests that the size of East Greenland polar bear sexual organs may be smaller with increasing levels of pollutants in the body. That in combination with climate change could pose a future risk to polar bear reproductive rates. A big geographical study of female and male polar bear genitalia size in relation to pollutants and climate change is recommended."

Overturning tortoises and the brain

OVERTURNED TORTOISES NEED TO RIGHT themselves as soon as possible. It's vital because when they are upside down they can have difficulty breathing and controlling body temperature, and if they stayed in that position for too long, they would perish either from lack of food, exposure of their soft parts to predators or the effects of sunlight on body temperature controls.

As tortoises have been around for so long, and the risk of being flipped over from fighting and traversing uneven terrain is high, they must have evolved a quick and safe way to get themselves back on their feet.

But how do they do it? How do they use their brain, and how long does it take the average tortoise to right itself?

To answer these and other questions, researchers at Padova and Trieste universities in Italy are spending time overturning tortoises and watching them right themselves.

The work is carried out in a special experimental plastic-sided sandpit. A wire mesh covers the sand surface to give the tortoises a better grip, and in the centre of the pit is a small rectangular depression into which the overturned tortoises are placed.

The attempts of the creatures to right themselves are then video-recorded and analysed.

Results so far reveal that there is a distinct shake-and-flip action universally used by upturned tortoises. It is, say the researchers, the first detailed description of such behaviour.

Detailed analysis has shown that there are two distinct movements involved. First, one leg is extended outwards so that it reaches and touches the ground on one side. Then, in order to unbalance itself, the head and all the limbs, except for the one touching the ground, are vigorously shaken. As a result of this rocking action, the tortoise turns right side up in the direction of the leg that was kept still.

The animals, which can live more than fifty years, are very efficient at the righting process. All tortoises tested so far have completed the righting response within two minutes.

The research, which has important implications for the development of the brain, and identifying cerebral structures involved in movement, also shows that most of the tortoises tested have a preference for righting themselves in a certain direction. Three times as many tortoises turned right to get themselves up as turned left.

That may give clues as how and why human brains have specialized functions. Brain laterality is all about how brains process information. Although the left and right hemispheres, or sides, of the brain are similar structures, they have specialized functions.

The left side, it's suggested, is more logic based and dominant and is the main language centre, while the right is more imaginative, more visual, intuitive, emotional and spatially aware. Because the right side of the brain controls the left side of the body, the left ear has been shown in some research to be the route to the emotional side of the brain, and the right ear to the non-emotional, logical side.

Counting worms

WORMS MAY BE GOOD FOR the garden, but they are bad news for golfers. That shot towards the fairway or green may start off being straight and true, but should it land and ricochet off a worm cast, the hole, even the game, could be lost. Too many casts can also cause damage to green-cutting machinery by blunting mower blades.

Most golf courses in the UK have had a strategy for worm control, but worms apparently have been raising their heads above ground in increasing numbers. Worm researchers have calculated that, in an average acre, worms bring 20 to 25 tonnes of soil to the surface each year in the the form of casts.

The problem is that in the old days, until 1998 that is, worms were kept under control with chemicals like Chlordane and highly toxic organophosphates that, once laid down, suppressed earthworm activity for at least seven years.

A total ban on use of these chemicals in the UK, from 1998, has resulted in greenkeepers reporting a substantial increase in problems with earthworms.

More environmentally sensitive ways of dealing with worms are needed, but the problem, as researchers and golf course owners and managers soon discovered, was that there was a dearth of data about worms. No one, for example, knew how many worms there were to the square foot, or what types, and whether some were more controllable, or more prone to making casts, than others.

That's why researchers at Cranfield University have been counting worms and casts on golf courses in England in the biggest study of its kind.

In the research, the academics first count the numbers of casts in defined sections of the fairways and then use a special technique, involving a solution made from mustard flour, to extract samples of worms from the ground over a twelve-month period. Worms that emerge within the first twenty minutes after mustard solution infiltrates the soil are collected, washed and stored prior to counting and identification. Earthworm populations are then calculated per square metre for each of the golf courses that have been tested.

Overall, seven species of earthworm have been identified as active by the researchers, including *Aporrectodea rosea* (Savigny), *Lumbricus rubellus* (Hoffmeister), *Aporrectodea longa* (Ude) and *Lumbricus terrestris*.

Two species of shallow-burrowing earthworms were most dominant (*A. rosea* and *L. rubellus*) and two deep-burrowing earthworms were present in significant numbers (*A. longa* and *L. terrestris*). Critically, at least from the golfer's point of view, all of these earthworm species maintain permanent or semi-permanent burrows that are open to the soil surface.

The researchers also found that earthworm populations differed markedly between the golf courses tested. The most abundant course had up to forty worms to the metre, compared with just two in the least dense.

The only significant difference that could explain the variations was the sand and silt content of the soil. The more sand and silt, the lower the number of worms, which may suggest a natural strategy for dealing with them.

"To golf course managers it is only the surface active earthworms that present any significant management problems," say the researchers. "This study indicates that the abundance and community structure of these species may now be significantly different from recent historic surveys or greenkeepers' traditional perceptions. The identification of the most dominant surface extractable earthworms means that research into environmentally benign controls for earthworm casting on amenity turf can be better informed in the post-Chlordane era."

Another finding of the research suggests that golfers who want to have the lowest likelihood of encountering a worm cast should aim to play the most games in July. That's the month when the fewest worms are about. The researchers found that worm populations can shrink tenfold during the summer months.

According to another worm researcher, Paul Backman of Washington State University, worms cast on the surface for two primary reasons. First, after they ingest organic matter, decaying leaf tissue and mineral soil, they must excrete the leftover material. Second, when soil fills the burrows, earthworms ingest the soil and move it up to the surface.

"Casting occurs when earthworms ingest soil and leaf tissue to extract nutrients, then emerge from their burrows to deposit the fecal matter, or casts, as mounds of soil on the turf surface," he says. "Extensive earthworm casting on fairways interferes with proper maintenance practices, the playability of the grass and the overall appearance of the fairways. Affected turf can become thin and the playing surface can soften."

Backman says earthworm populations can reach several million under golf course fairways, with millions more in the roughs going unnoticed owing to higher mowing heights. One of the problems he has identified is that the practices that produce excellent fairway surfaces also create ideal living conditions for earthworms, who feed on grass clippings returned after mowing and on organic matter in the soil.

The researcher, whose investigations began in 1998, has also identified *L. terrestris*, known as the night crawler, as the earthworm species causing severe casting damage on golf courses across the USA. He says they typically build permanent vertical burrows that vary in diameter from about 0.125 to 0.5 inch, and that they can be up to 12 feet deep. More bad news from the research is that night crawlers have long lives, with a reported average lifespan of six to nine years, and up to twenty in some cases. They are also rapid breeders, producing up to twenty offspring once every two weeks.

Meanwhile, worm researchers at Southern Illinois University have shown that worms may not be entirely bad news for golfers.

They have been working on the idea that letting loose an estimated three million earthworms on 70 tonnes of food that goes to waste each year will produce compost that could be used as a fertilizer on greens and fairways. The researchers have spent time calculating that in one year the university serves up 852,263 meals, which produce 172,157 pounds of food waste.

Researchers are now working out the amounts of compost to apply, and whether there are enough nutrients, and that could take some time: "We need to have at least two years of field data for anything that will hold up in terms of scientific review," they say. "Normally in agriculture research you want a minimum of two years' data for field work because of the kind of variations you get with rainfall, climate and temperatures."

Cow's teeth

STUDYING FOSSILIZED TEETH MAY NOT seem to be cutting-edge research, but it's helped scientists solve one riddle.

By analysing old fossils researchers have been able to show that the ancestors of the modern cow first appeared 9 million years ago, and that their shape evolved as a result of early climate change.

According to the fossil researchers, the early cows appeared in Asia and evolved a large body size and big teeth to adapt to grazing changes caused by climate warming at that time.

"This is the first report that the earliest members of the bovine lineage are present by at least 8.9 million years ago and that the origins appear to be in Asia south of the Himalayas," says Dr Faysal Bibi from Yale University. "I propose that the evolution of large body size was a consequence of climatic changes, namely the intensification of the dry season in southern Asia during the late Miocene."

The research centres on the analysis of fossils to try to trace and age the evolution of bovines, a group of animals that includes the African Cape buffalo, the American bison, the Asian water buffalo and yak, as well as the progenitor of domesticated cattle, the auroch.

The earlier evolutionary history of bovines is not known, and until now, the earliest bovines have been reported from around seven million years ago, from Africa.

The fossil analysis of material kept at Yale centres on new data extracted from dental material. That gives clues to both the age and evolution of the animals, and the change in the shape and function of the teeth over time.

The research report says that high-crowned teeth that date from around that time indicate reduced rainfall and local environmental aridification. Softer foods, it says, became unavailable and increased eating and chewing of coarse, fibrous vegetation became necessary for survival.

"The evolution of the large body size and robust dental characteristic of bovines was driven by climatic and environmental changes," says the report.

Meanwhile, cow-spotting researchers have tried to solve another question not on most people's lips – do cows prefer to lie on their left or right sides?

In theory, in an average herd, there should be equal numbers choosing each side. But does it work out like that in practice?

To investigate, researchers at the Swedish University of Agricultural Science spent some time counting the number of left- and right-lying cows in a number of experiments. Behaviour observations were made every twenty minutes in eighteen periods of four hours each in one study, and in a second, lactating cows were identified every twelve minutes during four periods of 48 hours each. In a third study, observations were randomly performed during the day for two months.

Results show that lactating cows choose equally between a left- and a right-lying side, although the researchers did find that among late-pregnancy cows, 60.7 per cent did favour the left side. This preference was probably due to the discomfort that is associated with lying on the right side caused by the foetus growing into the right abdominal cavity.

The sheep watchers

SHEEP MAY NOT BE SO dumb after all.

Research shows that they can choose the right medicinal food to eat when they are ill, as well as recognize and remember faces. When sheep were given food than made them unwell, they were able to select and eat the right cures for constipation and heartburn.

Research also shows that sheep can identify and remember faces, both humans and sheep, distinguish between happy and depressed expressions, and tell one sheep's bleating from another.

It's also claimed, but not yet proved beyond all doubt, that the animals have mastered the art of crossing cattle grids – by rolling over the bars or by having one volunteer lie down so the others can step over her.

In some of the latest research to show that the animals are not so woolly minded, sheep watchers at Utah State University have discovered that sick sheep can accurately self-medicate for stomach problems.

"People learn to take aspirin for headaches, antacids for stomach aches and ibuprofen to relieve pain, and often obtain prescriptions from doctors for medications. Is it also possible that herbivores write their own prescriptions?" say the researchers.

They claim that from prehistoric times people have believed that animals self-medicate, but that until now it has not been clear whether sheep can really spot medicinal compounds when they are ill.

In the research, lambs were given foods that led to mild ill health, and then given a choice of compounds including those known to ease the symptoms. The animals were able to accurately spot and eat the specific compound that would cure their ill.

"This is the first demonstration of medicine preferences in animals," say the researchers. "If we reason that animals are able to increase preference for substances that enhance homeostasis and states of well-being then it follows that self-medication is likely to occur in nature. Results from the present study support this notion."

Neuroscientist Professor Keith Kendrick, Gresham Professor of Physics at Cambridge University, says it all shows sheep are not as dumb as some thought. "We now have a fair amount of evidence that sheep are certainly not dumb. In fact, they can be quite cunning in terms of getting in and out of things, and coming back and looking as if they never went out in the first place," he says.

Professor Kendrick and his team have been investigating the face and emotion recognition among sheep.

"It is a review of how sheep process faces and emotions. We have found that sheep can recognize both human faces and emotions, and emotional changes on sheep faces. They are also able to form mental images of faces. They can recognize at least 50 different faces, and remember them for a couple of years or more," he says.

"They have the same kind of sophisticated way of recognizing faces as we do. They have a specialized part of the brain for doing it. They are quite sophisticated in their social environment. They know what a happy face looks like compared with an angry one."

Researchers at the French Behavioural Ecology Group have also found that ewes are able to recognize the individual sounds of their lambs, suggesting that sheep baas, which appear to the human ear to be all the same, may be unique to each individual.

"Our results show that ewes and their lambs can recognize each other based solely on their calls. These results point out a simple vocal signature system in sheep," they say.

Sheep facts from research:

- The average lifespan of a sheep is about seven years, but can be up to twenty.
- The largest sheep is the Argali of the Altai Mountains of Siberia and Mongolia, which can grow up to four feet tall at the shoulders.
- The sheep is believed to have been first domesticated in Asia during the Bronze Age.
- Sheep are warmer than humans with a body temperature of 102–103°F.
- Sheep bite the grass with their bottom front teeth.
- Lambs can walk within five minutes of being born.

Plight of the bumblebee

CHANGING FARMING PRACTICES HAVE PLAYED a key role in the decline of the bumblebee, according to researchers who have been counting the numbers of insects and found them to be falling.

The big drop in haymaking and the rise of silage are driving out the bees, whose numbers have declined by 60 per cent since 1970.

"We suggest that the widespread replacement of hay with silage, which results in earlier and more frequent mowing and a reduction in late summer wildflowers, has played a major role in bumblebee declines," say the researchers.

Once a common sight in the countryside in clover fields, hedgerows and around the edges of fields, most of the twenty or so species of bumblebee are in decline, and two have become extinct. Five bumblebee species are designated UK Biological Action Plan species, in recognition of their decline, with three more species scheduled for inclusion.

Exactly why there has been such a sharp decline is not clear, and there have been a number of theories, including climate change and intensive farming.

In the study, researchers point the finger at the decline in haymaking. Traditional hay meadows were not cut until after the flowers that attract bees had flowered. But with silage, grass is cut more frequently and early in the season so plants are unable to flower. More intensive farming means that the edges or margins of fields are now cultivated and are no longer havens for wildlife.

Silage is preferred by farmers for a number of reasons. There is less risk of weather damage and delays in harvesting, less crop loss, and the grass can be stored for long periods with little loss of nutrients.

In the study, researchers from Trinity College, Dublin, Queen's University, Belfast, Limerick University and the Ireland Institute of Technology looked at new data for Ireland, where more than half of the bumblebee species are in decline.

"We use a new and independent dataset based on Irish bumblebees. We found that most of the same bumblebee species are declining across the British Isles. We demonstrate that the late emerging species have declined in Ireland and in Britain and that these species show a statistically significant westward shift to the extremity of their range, probably as a result of changing land use," they say.

The research also shows that rare and declining bumblebees are the species that prefer open grassy habitats.

Listening to dogs barking

IN THE WORLD OF BARKING RESEARCH, scientists at Szent Istvan University in Hungary are pretty much top dogs.

By listening, recording and measuring canine chatter they have revealed hitherto unknown facts about barking. They have found, for example, that humans can detect different emotions in the barks of their dogs. They have shown too that the sounds of individual dogs have their own unique features and characteristics.

And in some of their latest groundbreaking work, they have discovered that the animals themselves can distinguish between the barks of other dogs in different situations.

With the help of heart monitors strapped to the dogs, the researchers were able to show that the animals reacted differently when they heard barks that had been recorded by other dogs when faced with a stranger, or when left alone.

"Dog barking has been considered a meaningless vocalization for decades. Recently we found that humans understand the emotional and referential content of dogs' barks. Now, with measuring the heart rate of dogs during testing, we showed that they can discriminate between dog barks, recorded in different situations," say the researchers.

For their researches, the team recruited a large and varied assortment of dogs, including German shepherds, Border collies, golden retrievers, German pointers, Labradors and mongrels, with an average age of five years.

The team also recorded the barking of a number of other dogs in two situations. The first was the barking that greeted a stranger entering the property where the dog lived, and the second was the barking of a dog tethered to a tree and left alone. Earlier human tests had shown that people were able to distinguish easily between the acoustics of these two sounds. Two other kinds of mechanical noise were also recorded for use for control purposes during the testing – the sound of a drilling machine on a brick wall, and the humming sound of a refrigerator motor.

Thirty minutes before experiments began in a special laboratory, the animals were partially shaved with an electric razor so that portable heart rate monitor wires could be attached next to the skin. Three electrocardiograph electrodes were placed on the skin of each dog to feed data to the monitor.

During each experiment, the owner sat on a chair in the laboratory, with the dog on its leash sitting in front. Both owner and dog faced the same direction. During the experiment, the owners were not allowed to touch their dogs.

The sounds were then played to the dogs in 25-second bursts and the reactions of their hearts were monitored and measured by the recorders.

One of the barking sounds was played a number of times so that the dogs got used to it. The second bark was then played to see whether they noticed any difference in the new sound as measured by their heart rate reaction.

The results show that the dogs reacted differently and were able to tell the difference between the barks. No such differences were detected for the two control sounds.

"Barks from the stranger and from the 'alone' situation have markedly different acoustic features, and our study here showed that dogs may have the capability to differentiate between them," say the researchers. "One possibility is that dogs can react and recognize dog vocalizations without any training. With measurement of the changes of the heart rate in dogs, we showed that dogs can differentiate between bark samples from two situations.

"This experiment has shown that dogs have the capacity for distinguishing between different barks. Our experiment showed that dogs can perceive the difference between barks originating from different situations, thus barking is perhaps a communicative tool not only for dogs to humans, but for dogs to dogs as well."

Grunting at baboons

NOT A LOT OF PEOPLE know this, but male baboons have evolved a way of achieving sneaky sex by eavesdropping on couples.

Bachelor baboons are able to home in on a female by listening to her grunts and those of her regular mate to seek how far they are apart.

If the distance is great enough, the unattached baboon takes this as a sign that mating opportunities might be possible, and moves in for what researchers describe as a sneaky mating.

All this may have passed everyone by, but for the work of baboon watchers at the University of Pennsylvania.

"Eavesdropping may be one strategy by which male baboons achieve sneaky matings," they say. "Male baboons monitor consortships so assiduously that they rapidly recognize both when an unexpected mating opportunity arises and when a consortship has ended."

It has been known that males often take over a female within minutes of another male moving out, suggesting that they are quick to observe opportunities. But until now just how the bachelor baboons spotted a chance to mate with a sexually receptive female has not been clear.

In a series of experiments, the researchers played different grunts and copulating noises through speakers to an audience of male baboons.

The consort male's grunts were played from one speaker and his female's copulation call played from a speaker approximately 40 metres away, suggesting that the male and female had temporarily separated. Sounds that suggested the pair were still together were also played. A third set of sounds, suggesting that the pair were apart, but that the female was already copulating with a third party, were played too.

Results show that the baboons were much more responsive to the sounds of the couple being apart.

"When male baboons heard a sequence of calls suggesting that a consorting male and his female had temporarily separated, they responded significantly more strongly," says the report of the research in the journal *Animal Behaviour*.

"They looked towards the speaker broadcasting the female's copulation calls significantly more times and also approached it. Their behaviour suggested that they inferred that the consort pair had temporarily separated, that the female was engaged in a sneaky mating, and that further mating opportunities might be possible."

The researchers say the results support the idea that baboons and other monkeys are able to recognize and understand social relationships. Unlike many humans, they may even be able to appreciate that some relationships just do not last.

"They seem to recognize that different types of relationships are characterized by different patterns of spatial proximity. Baboons apparently understand that some close social relationships are characterized by continuous proximity, while others are not. Moreover, they appear to recognize that some very close relationships are extremely transient," they add.

Avoiding crowds

COME THE SUMMER MONTHS, CAPTIVE gorillas could well be praying for rain. Summer brings out the crowds, and while the animals may be one of the big attractions, new research shows they don't like big holiday crowds.

Gorillas, especially the bachelors, are prone to shake their hands, thump their chests and clench their teeth when the crowds get big. They also show signs of stress, including picking, scratching or otherwise manipulating various parts of the body with hands, mouth or feet.

While the crowd flock to the gorilla areas, the attraction is apparently not mutual. When the crowds get big, the gorillas try to huddle out of sight as much as they can.

Two research teams have looked at the behaviour of gorillas when faced with large crowds. In the latest study, at Disney's Animal Kingdom in Florida, the gorillas were watched during holidays with big crowds and on other days when crowds were half the size.

Results show that gorillas were less visible when the crowds got bigger, and the bachelor group became more aggressive. Aggressiveness was defined as engaging in contact aggression, including biting and hitting, or non-contact aggression, including a stiff-legged stance, a tight-lipped face or chest-beating.

"Along with the growing concern for animal welfare in zoological parks, over the past several decades has come increasing interest in the impact of zoo visitors on the behaviour and welfare of zoo animals," say the researchers.

In a second study, at Queen's University, Belfast, researchers found that gorillas banged on the glass barrier seven times more often when the crowds were bigger. They also found increased aggression and abnormal behaviour when crowd sizes were large.

Just why gorillas dislike it when the crowds are large is not clear, but one theory is that it is it could be the size of humans, and the noise they make.

"Research conducted to date, albeit on other species of primate, implicates audience noise, activity and perceived height, but whether this holds true for captive gorillas is still unknown," said the Belfast researchers.

"The findings from this study suggest that gorillas, like many other primates, are excited by high numbers of visitors. Visitors can provide a unique and complex form of stimulation for many species of zoo animal. Nonetheless, captive-housed animals often find it difficult to escape the attention of, and disruption caused by, the general public," add the researchers.

Strewth, Joey, cover your eyes

THERE MAY BE MORE TO the sex life of the furry female koala than previously thought.

Research shows that the cuddly cousin of the wombat has an affinity for other females. When researchers trained digital cameras on them, they counted 43 homosexual and 15 heterosexual interactions.

"Some females rejected the advances of males that were in their enclosures, only to become willing participants in homosexual encounters immediately afterwards," say the researchers.

And worse: "On several occasions more than one pair of females shared the same pole and multiple females mounted each other simultaneously. At least one multiple encounter did involve five female koalas."

One theory is that they may do it to attract males, but the more likely is that it is a stress reliever.

The koalas, famous for eating little else but eucalyptus leaves, and for having offspring called joeys and unusually small brains, appear to develop the habit when they are in captivity, and only heterosexual activity has been observed in the wild.

In the research, scientists from the University of Queensland and other centres looked at the activities of the animals, including an analysis of the mating sounds, which include barking, bellows, soft groans and loud screams.

"The aim of the present study was to determine the extent of differences in the homosexual and heterosexual behaviour of female koalas and thereby to determine the purpose of female homosexual behaviour in the koala," says the report of the research with 130 Queensland koalas, which adds that homosexual interactions between females were spontaneous and not facilitated by staff.

"Wild koalas brought into captivity clearly display homosexual behaviour on a regular basis," say the researchers. "A total of 15 heterosexual matings and 43 homosexual interactions were recorded in separate animals on the digital camera used for recording. Homosexual behaviour was restricted to female koalas only, heterosexual encounters were typically twice as long as homosexual encounters."

The research found that homosexual behaviour in koalas was not usually spontaneous and normally occurred as a reaction to the presence of a male.

The researchers say there are several possible purposes of female homosexual behaviour in captive female koalas. They reject the idea that it could be a way of practising, because immature captive koalas were not seen to engage in such activity.

Another possibility is that it is done to attract males, but if it is, it was not successful: "Males that were present in enclosures when homosexual behaviour occurred did not always move toward the pair of females engaged in the activity, so mate attraction is probably not the objective of this behaviour."

Another nail in the coffin of this idea is the fact that the eyesight of koalas is poor.

Yet another theory is that it is hormonal and that the koalas are attracted to whoever happens to be handy: "It is likely, therefore, that the homosexual behaviour observed is an expression of a hard-wired behaviour that is stimulated in the hypothalamus in the presence of elevated oestradiol concentrations.

"In the presence of another koala, these behaviours are expressed whether the koala is female or male. If in the presence of a relatively dominant male then mating will occur, if in the presence of another oestrous female then homosexual behaviour will occur."

They add, "The behaviour may act as a stress releaser as it is only observed in captive populations."

TALKING TO PENGUINS, SPOTTING FAT BIRDS, AND RUDDY DUCKS

Dawn chorus

IT MAY SOUND PRETTY, BUT behind the trill of the dawn chorus lies a murky world of easy sex, multiple partners and male domination.

According to research on wrens, early morning birdsong is about male competition and older males wooing already partnered females.

One type of song – the chatter – keeps other males in their place, while the second – the trill – is a sign of male fitness, and is homed in on by females.

Or, as the researchers put it, "Both chatter and trill songs may have an ultimate role in gaining extra-group fertilizations."

Although birdsong before dawn has long intrigued researchers, its purpose remains poorly understood. Because many species sing at the same time, the dawn chorus has been put down to better acoustics at that time of day, or reduced risk of predators. It has also been suggested that it may herald the imminent arrival of the first food of the day.

In the research, evolutionary ecologists at the Australian National University looked in detail at the dawn chorus of the superb fairy-wren, recording the dawn singing of 36 adult males. Results showed that males began their dawn recitals 30 minutes before sunrise and produced between nine and 30 minutes of almost continuous song, interrupted only by occasional attacks on other males, and the odd extra-group copulation.

When they delved deeper, they found that there were the two types of song, chatter and trill. Observations showed that the chatter was more commonly used by the dominant male to let other males know he was around.

The trill, on the other hand, may be designed to attract females. The researchers found that the length of the trill increased with male age, which in turn was linked to attractiveness.

The fact that a male trills for a long time is a sign of good male quality. One theory is that older males trill for longer because they are in better condition, either because experience has made them good finders of food, or because they are better at handling stress.

"We do not understand the mechanism through which improvement occurs with age in fairy-wrens. Nonetheless, females may benefit from mating with older males because they have showed their survival ability or it may be useful for females to use a signal of male age in addition to one of quality," say the researchers.

They add, "The dawn chorus of the superb fairy-wren may thus have a dual role, involving enforcement of dominance among male group members, and signalling attractiveness to mates."

Counting birds

SPARROWS MAY NOT BE TOO CHIRPY about plans for more urban housing. According to researchers, house sparrows don't seem to like too many houses. They found that numbers of the birds, including chirpy males, decline rapidly when gardens and green spaces in towns and cities are converted to housing.

The bird-counting researchers found a threefold increase in numbers of the birds in areas where there were gardens, and they say green spaces need to be preserved to protect the bird.

"Allotments and residential areas with gardens are likely to be under pressure due to increased demand for housing, specifically from the infilling of green space within urban areas," say the researchers. "It would seem to be imperative that any action plan to protect urban house sparrow populations should include specific protection of such key habitats."

The house sparrow (*Passer domesticus*) is declining in many parts of Europe, with recent severe declines in urban areas. According to the Royal Society for the Protection of Birds, around 60 per cent of sparrows have been lost since the mid-1970s, and the declines have been particularly great in the south-east of England and in the centres of large cities, including London, Edinburgh and Glasgow. There are now thought to be between six and seven million breeding pairs.

The decline has been seen as particularly worrying because the house sparrow is one of very few species that actually thrives in close proximity to people, even in city centres.

It is also perplexing because, according to a report from the Department for Environment, Food and Rural Affairs, the number of birds in Wales and Scotland seems to be increasing, and rural houses and their gardens also support high densities of house sparrows: "Declines have been greatest in suburban and urban gardens. Rural gardens may be the most favoured habitat for the species," says the report.

Just why there has been such a dramatic drop in urban numbers is not known, although there have been a number of theories, including predation by domestic cats and sparrowhawks, loss of sources of weed seeds as a result of the development of brownfield sites, pollution and disease.

In the study, the researchers, who say little is known about the bird and its habitat in urban areas, counted sparrow densities in 1,223 randomly selected 500-by-500-metre urban areas in the UK where there was a relatively high human population.

The numbers of chirping male house sparrows and of all house sparrows were analysed separately, and the results showed that residential areas, allotments and farm buildings were key predictors of house sparrow density and chirping-male density.

"Within residential areas, the increase of house sparrow density with habitat area was approximately threefold greater when private gardens were present than when they were absent," says the report.

Results also show that there is a rapid decline in house sparrows when only a small area of private gardens is converted to housing.

Fear of cats may also be leading to a drop in urban birds such as the starling and sparrow, with research showing that Britain now has so many cats – more than nine million – that fear of them may be having as great an impact on the decline as the numbers killed.

While most research into the decline has centred on the killing of birds, the new study suggests that so-called sub-lethal effects or fear may be reducing bird numbers by affecting eating patterns and reproduction.

Especially vulnerable are songbirds, which are outnumbered by cats in some areas by up to 35 to one.

"We show that these sub-lethal or fear effects may be substantial for urban songbirds. When cat densities are as high as has been recorded in the UK, and even when predation mortality is low, a small reduction in fecundity [reproduction] due to sub-lethal effects, can result in marked decreases in bird abundances – up to 95 per cent," say the researchers.

The widespread decline of both rural and urban bird populations has been increasing for some time. Species such as the starling and house sparrow have declined by up to 60 per cent in urban areas of the UK over the past 30 years. Current theories include a reduction in insect food, loss of nesting sites, and predation by domestic cats.

In the new study, researchers says that the killing of birds by cats may not be the most important issue: "Predators influence prey populations not only by eating individuals, but also by altering behaviour, including foraging patterns and use of different habitats. It is emerging that the consequences of these effects may be larger than those of predation mortality."

The researchers looked at whether domestic cats had an effect on bird reproduction. The report, in *Animal Conservation*, says there are 9.2 million cats in the UK, an annual increase of 12 per cent over 40 years.

The findings may mean that while some strategies designed to stop cats killing birds may work, others, which make the animals more conspicuous, may not.

Talking to penguins

RESEARCHERS HAVE BRAVED FREEZING TEMPERATURES, icy seas and faked mating calls to answer one question – exactly how do penguins find their loved ones in a crowd?

Surrounded by hundreds, perhaps thousands, of near-identical black-and-white birds, all making loud noises, how do Magellanic penguins recognize their mates when they come ashore with food?

The answer, it seems, is that they home in on the calls of their mates and their chicks, and ignore the noises being made by the others.

"The results of these experiments strongly suggest that individual recognition through vocalizations is a key component of the breeding ecology of Magellanic penguins, for both adults and chicks," say the researchers from the University of Washington.

In the research, the calls made by individual penguins were recorded and then played back to see who recognized what and whom. Display calls, including those of a neighbour, stranger and mate, were played to incubating females and, later, to their mates, in a series of experiments.

The results showed that the birds were able to recognize their mate above the general background noise. That's important because the birds share parenting, and it's vital that they find their mate when they return with food.

The chicks also wander off when the parents are away, so it's equally important that they are able to recognize the sound of them returning. Chicks should be able to recognize parental calls and return to their nest; otherwise they may miss critical feeding opportunities.

Research results show that chicks, waiting for a parent's return to be fed, responded more strongly to their parents' calls than to calls of stranger pairs. Parental recognition by chicks is also important because chicks approaching adults that are not their parents may be subject to a physical attack.

"These results make sense ecologically, because an incubating female is expecting her mate to return to the nest to relieve her, and her mate's ecstatic display call, even if given at an unexpected time, announces her mate's return," say the researchers. "We demonstrate for the first time that adults and chicks can discriminate between calls."

The Magellanic penguin project was started in 1982 as a result of a Japanese company's intention to harvest penguins and turn them into golf gloves, meat and oil.

Based at the provincial reserve at Punta Tombo, Argentina, a small group of researchers led by Dr Dee Boersma, Professor of Biology at the University of Washington, follow individual penguins, monitor the colony and develop the data needed to plan effective conservation efforts, as well as to try to understand the importance of penguins as indicators of global climate change and the health of the environment.

For the last two decades Dr Boersma and her volunteers have identified penguins at Punta Tombo with name tags in the form of flipper bands. Each penguin gets a small metal band with a number on it, so as researchers walk the beaches recording data, they can tell who's who. The project has banded over fifty thousand birds since 1983.

Penguins are at the breeding colony during the spring and summer, from September to March, when researchers visit each nest and determine how each couple does raising their chicks. A few male birds, which have a reputation for being good 'dads', get a satellite tag so that where he travels when he searches for food can be tracked.

"By studying penguins while they are at sea we are able to learn where they go and why some birds are successful and others unsuccessful at rearing chicks," say the researchers.

Picking up penguins

THEY MAY BE SURROUNDED BY ice and snow, and they may look very, very cold, but penguins have their own form of central heating.

Scientists have discovered that their way of keeping warm is so effective it can boost sub-zero temperatures in Antarctica to the kind of balmy levels found in Dubai.

According to the research, the huddling-together habit of penguins raises temperatures as high as 37.5°C.

"By huddling together, emperor penguins generate a tropical environment in one of the coldest environments on earth," say the scientists.

The emperor penguin is the only bird that breeds during the severe Antarctic winter, and it is the males which have the job of incubation, which involves fasting for around 65 days. The problem is that successful breeding requires a temperature of around 35°C, and the temperature outside at that time of the year averages –17°C.

Faced with such low temperatures, no food to eat to boost energy levels, and with protective body fat fast disappearing, there is only one solution – huddling.

In the new research, scientists from a number of agencies in France, Japan and Australia, including the Université Louis Pasteur, the Australian Antarctic Division and the National Institute of Polar Research, investigated for the first time what exactly happens inside the huddles.

"The dynamics of huddling behaviour of breeding emperor penguins in their colony during their winter fast had never been studied," they say. "The total time birds spend huddling during their breeding cycle had never been investigated."

In the study, the scientists equipped a number of penguins in Adélie Land, which has around 2,500 incubating males during winter, with various data recorders. The aim was to see how long they huddled and how temperatures rose.

Results showed that the male bird spent an average of 38 per cent of the time huddling, with each individual huddle lasting around ninety minutes. The birds were able to move around so that each had equal access to the warmth of the huddles.

They were also able to control temperatures with a mix of loose and tight huddles: "As a consequence of tight huddles, ambient temperatures were above 20°C during 13 per cent of their huddling time. Ambient temperatures increased up to 37.5°C, close to birds' body temperature. In fact, the temperatures during 38 tight huddling bouts increased from 20°C to 37.5°C in less than two hours," say the researchers.

They add, "This complex social behaviour enables all breeders to get a regular and equal access to an environment which allows them to save energy and successfully incubate their eggs. Huddling behaviour of emperor penguins is a far more complex behaviour than previously described."

Out of 201 huddling bouts, as many as 17 per cent rose to an ambient temperature equal to or higher than 35°C.

Penguins during the pairing and incubation periods can adjust their exposed body surface areas by grouping loosely or tightly. This enables them to spend a large proportion of their time at temperatures above 0°C, while external temperatures are on average –17°C.

Who's a clever boy, then?

PARROTS, LONG RIDICULED AS MINDLESS MIMICS, can, it seems, add up, recognize shapes, solve problems, communicate in a similar way to very young children, and recognize optical illusions.

Research shows that the grey parrot, *Psittacus erithacus*, is no bird brain and has thinking abilities on a par with those of marine mammals such as dolphins, and apes.

Researchers say the findings have implications for the way parrots are treated in captivity, and that their newly discovered intellectual abilities need to be taken into account.

"Whether it is a zoo or in the home, people need to be aware that these birds need an incredible amount of intellectual stimulation and interaction. We have found that their communication skills are about that of a two year old, but that their adding and abilities with colours and shapes, more like a five or six year old," say the researchers from Brandeis University in the USA.

The oldest parrot tested was able to use English, to label 50 different objects, seven colours, five shapes and quantities up to and including six. He can identify, request, refuse and quantify about a hundred different objects, and can uses phrases like "Come here", "I want ..." and "Wanna go ..."

Researchers say the bird demonstrated number comprehension comparable to that of chimpanzees and very young children. Results of tests also show that he understands the concept of bigger and smaller, and same and different.

"It has become clear from this work that Alex exhibits cognitive capacities comparable to those of marine mammals, apes and sometimes four to six year old children. These results have demonstrated that Grey parrots can solve various cognitive tasks and acquire and use English speech in ways that often resemble those of very young children," says the researchers' report.

Results show that in various comprehension tests, the parrots were scoring greater than 80 per cent.

The research parrots, which are encouraged to interact with, and learn from, humans, are also now being tested for their ability to subtract as well as add.

"Although this work has interest on its own merit for researchers, it also has implications for the treatment of parrots in captive situations such as zoos or as human companion animals," say the academics. "We need to encourage an awareness of, and a sensitivity to, the abilities of non-humans, particularly non-primate and non-mammalian subjects. For far too long, animals in general, and birds in particular, have been denigrated and treated merely as creatures of instinct rather than sentient, intelligent beings."

They say the research demonstrates that parrots possess cognitive capacities that have until now been considered unique to humans and other primates. They say the long-term hope is that the material will be used to improve and enrich the lives of captive and companion animals.

Ruddy ducks

OVERSEXED AND OVER HERE ... Seven promiscuous, sexually aggressive ruddy ducks imported from America are responsible for the thousands of offspring that have now colonized more than twenty countries across Europe, threatening native wildlife including the endangered white-headed duck.

DNA tests have pointed the finger at ducks imported by the Slimbridge-based Wildfowl & Wetlands Trust in 1948. Some offspring later escaped, leading to a huge growth in numbers in the UK and elsewhere.

Researchers say the findings should spur the controversial plan to drastically reduce duck numbers.

There had been claims that many of the birds had arrived here naturally, like other American ducks, and so should be spared the cull designed to bring down the numbers of the colourful duck and save the white-headed duck.

But the new genetic evidence shows that the birds all have specific genetic material linking them to the Slimbridge seven.

"Our results confirm that the European ruddy duck population is likely to derive solely from the captive population in the UK and we find no evidence of recent arrivals from North America or of a mixture between ruddy ducks from Europe and North America," say the researchers, who report their findings in *Molecular Ecology*. "The genetic diversity of European ruddy ducks is consistent with a founder population as small as seven birds."

The researchers say that control measures have generated considerable controversy in the UK, and that the ruddy duck has been welcomed as an interesting addition by many birdwatchers.

"It is not surprising that opponents to control measures against ruddy ducks have argued that the growing population in Europe has been partly established by vagrants arriving through natural dispersal," say the researchers from Boston University.

To test whether the natural arrival of ruddy ducks from North America was involved, the team compared the genetic diversity of ruddy duck populations from both continents, including present-day birds at the Wildfowl & Wetlands Trust which are thought to be direct descendants of the Slimbridge seven.

The results show that there is considerably lower genetic diversity in the ruddy ducks from Europe compared with those from North America.

"Our results suggest that a small number of founders from North America, bearing only a fraction of the total genetic variability of the species, gave rise to a new population in Europe," say the researchers. "Seven ruddy ducks, four males and three females, were brought to Slimbridge in 1948 and approximately 90 descendants of these birds escaped between 1953 and 1973. These appear to be the founders of the present feral population across Europe and North Africa, as well as the birds still held in captivity in Europe."

They add, "Because the ruddy duck and the white-headed duck have been separate species for perhaps one to two million years, and because we can now rule out the natural arrival of ruddy ducks from North America to Europe, ongoing efforts to eliminate introduced ruddy ducks from European countries should be continued in order to conserve the white-headed duck."

Who's a pretty (big) boy, then?

WATCHING BIRDS EAT MAY BE BORING, but it can be productive.

Obesity among domestic budgerigars is a health problem and may be the result of high-energy diets, eating too much food, a lack of exercise, and a refusal to fly.

When researchers covertly filmed fat budgies, they found that the really plump birds walked to their food rather than flying. And the farther away the researchers placed the food, the fewer trips the birds made. But each time they ate more, so their calorie intake remained the same.

The only answer, say the researchers, may be to put dummy eggs and nesting boxes in the cages so the female birds, which are more prone to getting fat because males feed them as part of the courtship routine, use up more energy.

In their research report, academics from the universities of Berne and Groningen say that obesity in domestic budgies is a big issue: "Obesity is a common health problem in captive budgerigars. An increase in body mass is frequently seen in domesticated budgerigars. This may be caused by different factors such as a high-energy diet, high food intake, insufficient exercising, and genetics. Obesity may induce symptoms in birds comparable to the metabolic syndrome observed in obese humans," they say.

In the studies, the investigators experimented with putting food at different positions and heights to see whether this could reduce the intake. They used a number of different locations and filmed the birds. The images were then analysed for flight activity, food intake, behaviour and energy expenditure.

The research team say the welfare of budgerigars would be increased if a housing condition could be found that limited obesity. But the results showed that, rather than cutting back on food, the birds changed their eating behaviours to keep their energy expenditure the same.

"The feeding bouts were longer and the feeding frequencies were lower when the distance to the feeder was larger. The birds adjusted the frequencies and durations of their visits to the feeder in order to minimize their energy expenditure," the researchers say. "Another adaptation to the treatments was that the heaviest birds refused to fly to the resting perches and preferred climbing along the wire."

Breeding increases the activity of females, and the researchers say that one strategy may be to encourage females to lay eggs by providing nesting boxes.

They say it might reduce or even suppress female obesity, but add, "To avoid overcrowding, eggs could be removed and replaced by fake eggs. From a welfare perspective, the consequences of this procedure are unknown."

Another study at North Carolina University found that budgies on high-energy seed diet for 100 days put on more than 10 per cent in weight.

One of the problems, it seems, is that in the wild, where obesity doesn't happen, budgies are hard-wired to preserve energy. Flying requires eleven to twenty times more energy per minute than standing still, but captive budgerigars spent 94 per cent of their time on the perch.

Spotting fat birds

BRITAIN'S BLACKBIRDS ARE GETTING BIGGER, and it may be down to global warming.

Bird-spotting research shows that the birds are getting heavier by the year, and it's thought that it's due to the beneficial effects of climate change on the supply of worms at the right time of the year.

While other birds have been losing weight as a result of climate changes, the blackbird has been piling on the grams: "It seems that the blackbird has benefited from climate change more than other temperate species, possibly due to the changes in rainfall that have made earthworms, one of its main foods, more available," say the team of researchers from Cambridge University and a number of other centres.

In the research, reported in the journal *Oikos*, the zoologists investigated whether global warming has caused recent decreases in body weight and increases in wing length in several species of birds in the UK.

The results show that while, as predicted, there had been decreases in residual body weight in great tits, blue tits, bullfinches, reed warblers and blackcaps, there had been what is described as a surprising increase in the body weight of blackbirds.

The researchers say the reason for the increases may lie in the different ways that birds have reacted to global climate change.

"Short-term variation in temperature had little effect, but rainfall did explain the unusual increase in blackbird body weight, possibly as a result of improving food – earthworm – availability," say the researchers. "The significant positive relationship found here between blackbird body weight and annual rainfall supports this hypothesis."

It's suggested that the weight gain may give the blackbird the edge in competition with other birds, and that could be bad news for other species, which have already seen the blackbird expand its territory.

Before 1850, the blackbird was considered a shy woodland bird, but by the 1930s it had spread to rural and suburban gardens throughout the UK and western Europe, as well as Finland and Sweden.

Sexual fetishism in a quail

MANY PEOPLE HAVE SEXUAL FETISHES. It could be a liking for leather, rubber, PVC, paper, or just about anything.

But how are they acquired, what parts of the brain are involved, and more importantly, if unwanted, how can they be got rid of?

The dominant theory is that fetishism is down to behavioural conditioning, the theory that objects and materials of almost any description can become erotic if they are associated, usually early in life, with sex.

One of the frustrations for researchers in this area is that, for obvious ethical reasons, experimental studies of human sexual conditioning are limited to individuals who have reached legal age, by which time their particular sexual preferences are usually well established.

Experiments have exposed college students to pictures of neutral objects, such as shoes or coins, intermixed with heterosexually erotic pictures, but they did not develop a compulsive fetish. This is likely to be because the students were sexually mature.

So to answer these and other questions, researchers at the universities of Istanbul and Texas turned to the quail. Their aim was to see whether they could condition the birds so as to induce fetishism in the male quail, and the fetish they chose was a sexual liking for nappy or terrycloth material.

A sexual fetish is defined as use of a non-living object as the exclusive or preferred method of achieving sexual gratification. It has to persist for a minimum of six months, with "recurrent, intense sexually arousing fantasies, sexual urges or behaviours involving the use of non-living objects".

Sexual interaction with the object leading to sexual gratification is an important feature of fetishism. Exposure to the object may be necessary for, or may facilitate, successful sexual interactions with a partner.

The aim of the researchers was to condition the males to mate with a terrycloth object by getting it to associate the terrycloth with sexual activity.

The conditioning procedure for male quails used by the researchers involves presenting the terrycloth to the male shortly before allowing it to copulate with a female. Copulation in the quail begins with the male grabbing the back of the female's head, mounting on her back with both feet, and then bringing its cloaca in contact with the female.

The theory is that by being first exposed to a female and then terrycloth, which is in the shape of a soft object that the bird can grab and mount, the male will associate the material with sex and will acquire a liking for the terrycloth alone.

In the research, the academics had one group exposed to a female and a terrycloth, and another group to a female and a flashing light. The conditioning stimulus was therefore either a terrycloth object filled with soft polyester fibre that the male was able to copulate with, or the flashing light.

Results show that half the males acquired a fetish for copulating with the inanimate terrycloth object. In fact, they kept on copulating long after the female had gone. None was turned on by the light.

"Several aspects of the findings support the contention that conditioned copulation with the terrycloth object provides a good animal model of fetishism," say the researchers. "First, the sexual arousal was elicited by an inanimate object which is one of the defining characteristics of a sexual fetish. Second, the behaviour persisted in the absence of copulation with a female, another defining characteristic of fetishism."

They add, "Such observations suggest that the terrycloth became a surrogate sexual object, in a manner analogous to human examples of fetishism."

The researchers say their findings are important for future research. They may help with the study of genetic and biological factors involved in fetishism. Knowledge of neural mechanisms may, they say, be particularly useful in understanding the development of sexual fetishes.

In a second experiment the males with and without a fetish for terrycloth were then investigated for reproductive performance with a female quail, whose eggs were then incubated to assess rates of fertilization.

Results showed that the birds which had acquired a fetish were slower to achieve sexual contact with the female quail and showed less efficient copulatory behaviour. But they also fertilized a greater proportion of eggs than non-fetishistic males.

5

CLIFF RICHARD, TALKING TO WITCHES, THE QUEEN, AND BRITISH PRIDE

British pride

DECLINING NUMBERS OF PEOPLE ARE PROUD to be British.

People who are very proud of their Britishness are now in a minority for the first time, and the number of people who are not proud of their country has doubled in 25 years.

Only 45 per cent now class themselves as very proud to be British, compared with 60 per cent in the early eighties, with the biggest drops in pride in Scotland and Wales.

The decline of the empire and religion, and increased urbanization and cosmopolitanism, are among the reasons put forward for the decline.

Oxford University researchers who carried out the study say that as older generations with stronger British national identities die out there will be greater potential for independence movements to make headway.

"We have shown that the decline in the numbers of people proud to be British is real. The average person is clearly less proud of their Britishness than the average person was in 1980," they say.

"The major driver of this decline of pride in Britain is not change in the composition of the British population but rather a generational shift in sentiments.

"In some ways there is bound to be a continuation of the decrease because older generations who are very proud are going to die and they are being replaced by younger generations less proud," they go on.

"There are a number of theories for the decline we have seen. The source of pride in Britishness was linked to Empire and Britain's international role, and clearly that has changed. There have been social changes too, including increased cosmopolitanism which perhaps makes national identities less important."

In the study, the academics used data from a variety of sources to trace national pride in Britain back from the present day to the beginning of the 1980s.

Results show that the number of people classed as very proud has dropped from a peak of 61 per cent to 45 per cent over the study period. The number of people saying they are somewhat proud has increased from 33 to 41 per cent. The number not very proud went up from a low of 6 per cent to 11 per cent.

There were differences by social group with Protestants, non-graduates, the working class and the petty bourgeoisie being the most proud of their country. Results also show that Scottish people are much less proud of their Britishness than are English or Welsh residents.

The authors say that younger people have grown up in a world of more complex identities, in which Britain is less prominent internationally and British identity in a multicultural, multinational state is less distinct and more frequently called into question.

"They are thus less likely to have acquired the strong attachments to Britain that older generations acquired in their youth and maintained throughout their adult life," they say.

The researchers also found that the decline was less during the early Thatcher years: "The stalling of generational decline in national pride among English people coming of age during the 1980s, when the dominant political party was broadcasting an assertive, and strongly English, brand of nationalism, is also highly consistent with explanations that relate to formative experiences, especially the political environment in which one is raised."

They point out that both Wales and Scotland have also seen the emergence of nationalist movements that provide a challenge to British identity and sentiment, whereas England has not as yet seen the emergence of a specifically English movement.

"Younger generations elsewhere on the other hand, especially in Scotland, appear to have been particularly receptive to the appeals of nationalist movements that have provided alternative foci for loyalty and led to even greater disparities between generations in a sense of British national identity. This can also be seen in the patterns of different types of British pride that newer generations of Scots tend to be less receptive to," they say.

The authors warn, "These changes, as with all generational changes, have inevitable implications for the future. As older generations with stronger British national identities die out, the 'glue' holding the different parts of the UK together is likely to become weaker still.

"The success of such movements will depend on political contingencies that cannot be predicted, but it appears from the evidence we present here that affective attachments to Britain will provide a weaker defence against separatism than they have done in the past."

Talking to witches

INCREASING NUMBERS OF TEENAGE GIRLS and young women are, it seems, identifying themselves as witches.

According to researchers, sales of spell books and other paraphernalia are booming, and they found 700,000 Internet websites aimed at or about teenage witches

"There has been a noticeable rise in the number of young people identifying themselves as witches. The author observed this amongst university students, including those training to be teachers," say the researchers. "The Pagan Federation claims to have several hundred enquiries a week from young people, and has set up a network for those under 18."

According to the report, based on interviews with young witches, a main attraction is witchcraft's attitudes to women: "Paganism and witchcraft appealed because of their clear feminist credentials and absence of homophobia. They agreed a main attraction is the positive valuation of women in comparison with other religions," says the report by Professor Denise Cush, head of religious studies at Bath Spa University.

It says that in the last ten years there has been an explosion of easily available material about witchcraft, much of it aimed at teenagers:

"The influence of the Internet in spreading teenage witchcraft cannot be underestimated," it says. "Much of the material is aimed at teenage girls rather than boys. The 'teen witch' phenomenon is largely, but not exclusively, female."

It describes how the material available includes spell books that focus on gaining money or career success, or even compelling other people's partners to prefer the practitioner.

"It does not seem to be an accident that the increase in teenage witchcraft has coincided with an increase in the availability of easily available information, whether books or Internet sites," claims the report.

"They mostly practised on their own rather than as part of an organised group, and they did use rituals as a form of self-empowerment and to help cope with events like examinations, friendships and common anxieties of young women," the report goes on.

"They all saw what they variously described as being a witch or being a 'Pagan' as a serious religious path which they had been following for many years since their early teens, and an important expression of their identity.

"Their motivations did not seem to be the selfish and trivial ones of white witchcraft but among the attractions of their chosen religious path were a concern for the natural environment and social justice. Being a witch was an expression of an alternative and special identity."

Cliff-hanger

AT FIRST SIGHT, POP SINGER Cliff Richard may seem an unlikely subject for an academic study.

But sociologist and writer Anja Löbert, who is described as the first international Cliff Richard researcher, and who is affiliated with the Martin Luther University of Halle-Wittenberg in Germany, has found there's more to him that may seem apparent.

Her intensive studies have concluded that the veteran singer presents himself as a redeemer and saviour whose similarity to Jesus Christ is difficult to ignore, according to the first academic study of the work and lifestyle of the iconic singer.

Song lyrics, photographs, lighting and stage performances where he is elevated towards heaven looking down on upturned faces, or portrayed in a biblical pose, all reinforce the image, says the report.

His permanently young, death-defying appearance, incorruptible body, apparent asexuality, his charity work for the sick and poor, and his white suits and lack of sex appeal also play their part in the saviour image, it says.

"Cliff Richard seems to have established a remarkably consistent public persona which presents him as a friend in need, a bridge to happiness and, hence, a figure modelling itself on Jesus Christ," says her report in the Cambridge University Press journal *Popular Music*. "The benevolence, apparent incorruptibility, and asexuality of his public image in combination with his lyric themes, certain habits of posture, and photographic tricks with light and with elevation, lead quite easily towards an interpretation of Richard as a redeemer figure. The relationship between him and his fans appears to rely to a considerable degree on the myth of redemption which the singer offers and the fan seeks."

In the study, Anja Löbert analysed song lyrics, official picture releases, video footage and other resources, and interviewed fans.

Analysis of song lyrics, says the report, shows repeated themes of redemption, incorruptibility and undying love, which presents the singer as a redeeming friend with an ever-understanding, ever-loving, trustworthy character.

The report questions the repeated use of the first person singular in lyrics – as in "Let Me Be The One" – and points out that other singers use alternatives.

"It is [unclear whether] Richard is merely pursuing missionary endeavours as a Christian by means of this poetry, or whether he is, in fact, presenting himself as the Saviour," says the report.

It adds that there are a number of official images of the star which have religious connotations. Of one it says, "Both arms are spread out at an angle of 110 degrees from the body and reach out towards the sun. The palms of his hands are opened, the face entirely bathed in sunlight. The figure is photographed from a low angle camera perspective through a transparent sky."

It says that in other images he stretches out his arms like the crucified Jesus. Some images, it says, are similar to the Rio de Janeiro monument of "Christ, the Redeemer".

Of the sleeve image for the "Let Me Be The One" CD, the report says, "The visual effect makes the singer appear to be a figure located in the sky – a heavenly figure. The low position of the camera emphasises the supremacy and power of the figure. This actual and symbolic body-enlargement is to be found on numerous occasions in Cliff Richard's self-presentation."

The report describes how, during the performance of the song "Remember Me", the singer, dressed all in white, is standing on a small platform that gradually arises heavenwards as he promises the upturned faces: "Remember me, I am your guardian angel, and I'll never let you fall."

The report also cites the pop star's charitable works over the last 30 years, especially in the Third World, including a visit to Mother Teresa's Home for the Destitute and Dying in Calcutta.

It points out too that just as Jesus Christ has been depicted as overpowering death, Cliff appears to have overcome the natural process of ageing.

"The denial of death which Cliff Richard seems to personify can be seen as one factor in the whole of the semantic field that suggests the singer's reading as a Jesus-figure," says the report.

It adds, "Another relevant component of Richard's image is that of his sexual inoffensiveness. Not only is he an unmarried Christian fundamentalist with a decisively celibate lifestyle, but in contrast to singers like Mick Jagger, Tom Jones, and Elvis Presley, he is normally not associated with sexiness.

"The same characteristic of sexual inoffensiveness can be discovered in his singing voice: controlled, with an absolutely professional check on it, soft, gentle, and in every sense conveying comforting warmth rather than unsettling anxiety, cleansed purity rather than guilt-laden sensual indulgence.

"The semantic proximity to Jesus is evident: he remained unmarried and is normally neither represented nor perceived as a sexually desiring or desirable being."

Just what Sir Cliff will make of it is not known, but it's possible that he might let another of his most famous lyrics speak for him ... "But I'm not like that at all".

Name games

DO YOU FAVOUR ARMANI AND ADIDAS as designer labels, and have you ever been an accountant?

If so, it's more than likely your own name includes the letter A and that you are a casualty of name letter syndrome, or NLS.

According to researchers, people are attracted to brands, careers, cities and even wives and husbands whose names begin with one or more of their own initials.

"The name letter effect is the tendency to evaluate alphabetical letters in one's own name – especially first and last initials – particularly favourably. Objects that match initials might trigger a sense of ownership just like the initials themselves," says Dr Gordon Hodson, who led an international study which included 150 people from the UK.

And it's not just fashion labels. Research shows that even careers and homes can be chosen as a result of NLS. Researchers found that more roofers have initials with an R, and more hardware shop owners H initials, than would normally be expected. They also found a greater than expected number of people called Louis living in St Louis.

Yet another study shows that both men and women prefer to marry people who share their last-name initial.

In research at Western Ontario and Brock universities in Canada, psychologists homed in on whether or not people were attracted to brand names as a result of NLS. Men and women were given the task of ranking 26 brand names, including Honda, Nike and Timex, each beginning with a different letter of the alphabet. A nine-point rating scale was used, and at the end of the exercise the numbers of times each person's initials appeared on the most-favoured list was counted.

"The name letter effect emerged as did a clear preference for brand names starting with one's name. The fact that such subtle findings can be found for marriage partners and brand name preferences suggests the powerful influence of name initials," says Dr Hodson.

The same name effect has been found by other researchers. Academics at the University of California investigated the effect of name resemblance on strikeouts in baseball. Strikeouts are recorded using the letter K, and after analysing 93 years of baseball results, they found that batters whose names began with K struck out at a higher rate than the remaining batters.

In another study, they looked at grades given to students in academia. Based on fifteen years of data on graduates, they found that students whose names began with C or D earned lower grades than those whose names started with an A or a B.

"Students with the initial 'C' or 'D,' presumably because of an unconscious fondness for these letters, were slightly less successful at achieving their conscious academic goals," says a report on the research.

Another finding along the same lines was that as the quality of law schools declined, so too did the proportion of lawyers with name initials A and B.

Researchers at the University of Buffalo have found name effects in the cities where people choose to live. One study they report showed that people disproportionately inhabit cities whose names feature the numbers in their own birthdate.

"Just as people born on February 2 (02–02) disproportionately inhabit cities with names such as Two Harbors, people born on May 5 (05–05) disproportionately inhabit cities with names such as Five Points," according to the report.

"Other studies showed that self-associations like name letters guide people's preference for streets and states of residence. Women named Louise, for example, are disproportionately likely to live in Louisiana, even if they weren't born there. People are disproportionately likely to marry others who happen to share their first or last initial."

In another name-counting study at the London School of Economics, researchers looked at the alphabetical ranking of names on academic papers to see whether this had any effect on reputation. They found that faculty members with earlier last-name initials were more likely to get employment at high-standard research departments.

They also found that the advantage of being listed as the first author motivates academics to manipulate their names so as to obtain a more beneficial alphabetical position.

Probing the brains of taxi drivers

BECOMING A CABBY MAKES YOUR BRAIN, if not your head, get bigger.

When scientists compared the brains of London cabbies and busmen, they found that the taxi drivers had more grey matter in the area of the brain associated with memory.

They believe that this part of the brain, the mid-posterior hippocampus, is where cabbies store a mental map of London, including up to 25,000 street names, and the location of all the major tourist attractions.

The research is the first to show that the brains of adults can grow in response to specialist use. It has been known that areas of the brains of children can grow when they learn music or a language, but it has not been shown to occur in adults before.

In the study, researchers at the Wellcome Trust Centre for Neuro-imaging at the Institute of Neurology, University College London, carried out scans on the brains of 35 cabbies and bus drivers, all men. Various psychological tests were carried out too.

Using bus drivers meant that any brain differences found could not be explained by driving stress, or dealing with passengers and traffic in London. The one big difference between the two is that the bus drivers stick to routes, while cabbies have to learn the layout of 25,000 streets in the city, and the location of thousands of places of interest, in order to get an operating licence.

The results of the scans show that the mid-posterior hippocampus of all the cabbies was bigger and that they had more grey matter than the bus drivers.

"We have found that London taxi drivers have greater grey matter volume in the mid-posterior hippocampi and are better at identifying London landmarks and knowing their position than London bus drivers," say the researchers. "Licensed London taxi drivers show that humans have an amazing capacity to acquire and use for navigation a highly complex spatial representation of a large city. Our findings suggest that this is accompanied by greater grey matter volume in the posterior hippocampus."

They found that in another area of the brain, there was a small drop in grey matter, but the significance of that is not clear.

Dr Eleanor Maguire, who led the research, said, "It is an interesting study because it compared two types of professional driver operating in the same area and exposed to the same kind of traffic and driving stress. Despite those similarities, we found significant differences in that area of the brain.

"We speculate that this may be the area where they store the mental map of London. We know this part of the brain is very important for navigation. What is unique is that we are seeing these changes in adulthood, and that may have important implications. We are now looking at the brains of taxi drivers before they start training, and at those of retired cabbies to see whether the area of the brain gets smaller when it is not used."

But the scientists warn that the arrival of GPS could change all that. New cabbies who do not learn routes and landmarks and rely on GPS instead may find that this area of their brain remains the same as that of mere mortals.

"GPS may have a big effect. We very much hope they don't start using it. We believe this area of the brain increased in grey matter volume because of the huge amount of data they have to memorize," says Dr Maguire.

"If they all start using GPS, that knowledge base will be less and will possibly affect the brain changes we are seeing."

Football strikers really are fit

IN FACT, THEY AND THE GOALKEEPER are the best lookers in the team ... and scientists reckon they know why.

Research based on the faces of professional players shows that women consistently rate the strikers and the 'keepers as being much more attractive than the rest of the team, especially defenders. According to scientists, it's because their faces reveal athletic flair. Attractive facial looks, it's argued, are part of an inherited package, along with the particular fitness skills needed for playing in those two positions.

In the research, more than seventy women rated pictures of the faces of professional football players. In cases where women recognized one of the players from media pictures, their rating was not taken into account.

"The key results are that the faces of strikers and goalkeepers were judged to be more attractive than the faces of defenders," say the researchers from the Royal Netherlands Academy of Arts and Sciences and the University of Groningen.

The researchers say the results are consistent with the theory that, because of differing physical and psychological demands, the positions of goalkeeper and striker depend more strongly on traits associated with inherited fitness.

Because the main job of 'keepers is to prevent goals being scored, they need to be especially quick, agile and willing to risk injury, say the researchers. And because the primary task of strikers is to overcome the defence and score goals, they need to be spontaneous and creative, and require more explosive strength. Defenders, on the other hand, require high levels of endurance, but less agility, spontaneity and creativity.

The theory is that the inherited traits of agility, bravery, creativity and spontaneity are more likely to come with another inherited sign of fitness – good looks.

"The results are consistent with the hypothesis that, because of differing physical and psychological demands, the positions of goalkeepers and strikers depend more strongly on traits associated with heritable fitness," say the researchers, who describe the results as remarkable.

The researchers carried out a similar study with professional ice hockey players and found exactly the same results.

They say the results pose a number of questions for future research. Are, for example, 'keepers and strikers more symmetrical than their teammates, because symmetry is also a sign of fitness?

And a more intriguing one: "Do they emit more sexually attractive odours, especially for women in the fertile phase of the monthly cycle?"

We speak the Queen's English, we does

LISTENING TO THE QUEEN'S CHRISTMAS MESSAGE may be a considered a chore for many, but for Professor Jonathan Harrington it's almost become a way of life.

He analyses her speeches not so much for what Her Majesty says, as for how she says it, and in particular her use of vowels. And one of his findings is that she has started speaking with a hint of Estuary English, an accent that is a mixture of standard BBC English and cockney.

Buckingham Palace may be outside the range of Bow Bells, but according to the research, the queen has subtly shifted some of her vowels in the direction of Estuary English.

His painstaking research, analysing 50 years of the queen's annual Christmas broadcasts, shows that over the last half-century her vowels have changed. In the research, the audio broadcasts were analysed individually by his team, digitized, and compared. They homed in on vowels in particular to see whether there had been changes over half a century.

The "happy Christmas" favoured in 1950 has become "happee", "dutay" is now "dutee", and "one's hame" is now "our home".

"There has been a shift in these vowel sounds towards Estuary English. Usually these kinds of sound changes are initiated by the young, and a person's accent is thought to have become stable by adulthood. But these results show this is not the case and that the Queen has changed her accent in the light of the changes that are taking place in the wider community," says Professor Harrington, who holds the chair of phonetics at the University of Munich.

"These changes have also occurred in Estuary English which has been influenced by cockney. Obviously the Queen's accent has not become cockneyfied but it has shifted subtly towards an accent that is more typically spoken in the wider community.

"The changes also reflect the changing class structure over the last 50 years. In the 1950s, there was then a much sharper distinction between the classes as well as accents that typified them. Since then, the class boundaries have become more blurred, and so have the accents. Fifty years ago the idea that the Queen's English could be influenced by cockney would have been unthinkable.

"The most important conclusion is that adults' accents do change over time under influences from the community, and there is very little that anyone can do about it," concludes Professor Harrington.

The results of the studies also show that the average broadcast lasts 5 minutes, 36 seconds. The shortest was in 1959 (61 seconds) and the longest in 1956 (8 minutes, 21 seconds).

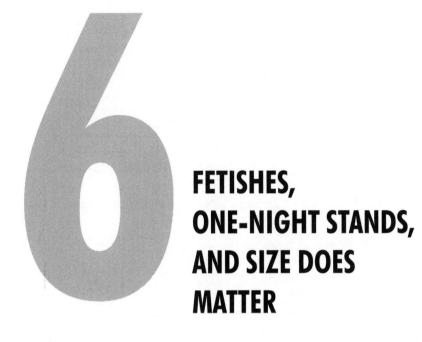

FETISHES, ONE-NIGHT STANDS, AND SIZE DOES MATTER

Making sex five times better

IT MAY SEEM OBVIOUS THAT SEX with someone else is better than being a solo practitioner, but how to measure the difference, and exactly how much better is it with a partner? And why is intercourse health-promoting, and masturbation not?

To answer these and other seemingly vital questions, researchers have been recording data on volunteers who were tested as they had sex, either with a partner or alone.

While sex is under way in a laboratory, an intravenous cannula or thin tube previously inserted into a vein in the upper arm of the participants takes a small sample of blood every minute for the next hour. The blood is removed with the help of a pump, which is located in an adjacent room in order not to distract the volunteers from the task in hand.

All the volunteers in the research at the Hanover Medical School are first checked to ensure they are comfortable with the idea of sexual activity in the laboratory, and then divided into those either having sex with someone else or masturbating, or a control group who get to do neither.

In the masturbation group, the experiment involved lone sexual activity in private while watching an erotic film. In the intercourse experiments, the couples viewed an erotic film in the lab together. The control condition, for those unlucky enough to draw the short straw, involved sitting silently watching a non-sexual documentary film without any form of physical contact with their partners.

The men and women volunteers are monitored from a neighbouring room by the researchers and are encouraged to shout via an intercom the moment orgasm occurs so that it can be recorded on the monitoring data.

During the activities, the researchers analyse the blood flowing from the next room, looking in particular for a hormone called prolactin. High levels of it are known to diminish sex drive, and research has shown that it is found at higher levels in women while they are breast-feeding, and that men and women with prolactinoma, cancer of the pituitary gland that results in greatly increased output of the hormone, have little or no sex drive.

Prolactin also kills sex drive after orgasm. It is released only after orgasmic sex and is thought to work through the dopamine brain circuits to dampen down desire. Research has also suggested that the more pleasing and satisfying the sex, the greater are the levels of the hormone released.

One of the objects of the research was to see whether solo sex or sex with a partner produced the more prolactin. The theory is that the amount of post-orgasmic prolactin is a marker of how good the sex was. The higher the level, the greater the satisfaction.

Results of the experiments show that for both sexes intercourse produced a substantially greater post-orgasmic prolactin increase than did masturbation. In fact it was five times higher.

One theory is that the complex physical and emotional changes that occur during intercourse cause changes in brain chemicals, including dopamine, which gives rise to the increase in levels of prolactin.

It has been known for some time that regular intercourse is health-promoting, but just why has not been clear. In healthy young adults, for example, it has been shown that frequency of intercourse, but not of other sexual activities, is associated with better cardiovascular autonomic functioning.

Women who have intercourse at least three times a month have better vaginal health than women who have it ten times a year. Research also shows that, in men, daily blood levels of prolactin increased with higher rates of intercourse.

The researchers say their findings raise the possibility that at least part of the mechanism by which intercourse leads to a better physical and mental health may be dependent on the release of greater amounts of prolactin.

While the Hanover team have been looking at making sex better, other researchers have been delving into why we have sex, and found that there are 237 reasons.

While love and attraction are the top clincher for many, for others it's about getting closer to God, gaining a promotion, revenge, or a way of getting rid of a headache. Some see it as a reasonably effective way of overcoming boredom, or burning up calories, while a few were attracted by the idea that it kept them warm, helped them fall asleep, or eased the stress of the day.

Results of the biggest study carried out on sex motivation show that men are much more likely than women to be triggered into action by looks.

"We identified 237 distinct reasons why people have sex," say the researchers. "The current research provided perhaps the most comprehensive exploration to date of the reasons people express for having sexual intercourse."

In the study, psychologists at the University of Texas quizzed more than two thousand men and women aged 17 to 52 about the reasons why they or someone they knew had sex.

"Why people have sex is an extremely important but surprisingly little studied topic. One reason for its relative neglect is that scientists might simply assume that the answers are obvious – to experience sexual pleasure, to relieve sexual tension, or to reproduce," say the researchers.

While being attracted to the other person was the main reason for both men and women, the results show that for some people the reasons were not so mainstream. They show that each of the 237 reasons was given the highest rating by at least one of the people taking part in the study.

"The frequently endorsed reasons for having sex reflect what motivates most people most of the time – attraction, pleasure, affection, love, romance, emotional closeness, arousal, the desire to please, adventure, excitement, experience, connection, celebration, curiosity, and opportunity," say the psychologists.

"One person's seemingly trivial reason for having sex might well be another person's magnificent obsession. For nearly every reason, some individuals gave it the highest rating, indicating that it was the most frequent reason for having sex in their lives."

Fetishes

FEET AND SHOES ARE NOT just for walking after all. The feet, toes and objects worn on them are the most popular fetish. Forget legs, buttocks, breasts and thighs – it's feet which are setting hearts racing.

Results from the biggest and longest study of sexual fetishes show that feet have seven times as many followers as hair, while footwear, including shoes and socks, sent the pulses of three times as many people racing as did a glimpse of underwear.

"Feet and objects associated with feet were the most common target of preferences," say the researchers. "Feet and toes, as well as socks and shoes, received most of the preferences. This is the first large survey on the relative prevalence of unusual sexual stimuli on a very large worldwide sample of people interested in fetishism."

The international survey, based on the preferences of both men and women, also reveals some of the more obscure attractions, including 150 people with a peccadillo for hearing aids, and two whose hearts were put into overdrive at the thought of pacemakers.

The researchers, who have published their extensive work in the *International Journal of Impotence*, say little is known about fetishes, and that most existing data has been based on clinical cases with psychiatric patients, sex offenders and people in therapy.

The aim of the study was to survey sexual preferences in the general population rather than clinical cases of fetishism.

"In everyday usage, fetish refers to sexually arousing stimuli that would not meet psychiatric criteria for a diagnosis of fetishism. In many cases, they may simply enhance sexual interest or satisfaction rather than being necessary for it," say the researchers from the University of Bologna.

"The word fetish that we used to locate data on sexual preferences is used in everyday language with a much broader scope than its psychiatric definition, and the two should not be confused."

In the study the psychologists monitored activity in discussion groups on the Internet dedicated to fetishes. They say that the groups monitored have up to 150,000 members, although the number of people surveyed was probably around five thousand.

The researchers estimated a relative frequency rate for each fetish based on the number of groups devoted to the category, the number of members, and the total number of monthly messages.

The results show that preferences for body parts or features and for objects usually associated with the body were most common (33 and 30 per cent), followed by preferences for other people's behaviour (18 per cent), own behaviour (7 per cent), social behaviour (7 per cent) and objects unrelated to the body (5 per cent).

The results for sexual preferences for body parts show that feet and toes had a frequency of 47 per cent, compared with 9 per cent for both body fluids and body size, 7 per cent for hair, 5 per cent for muscles, 4 per cent for body modifications such as tattooing, 4 per cent for genitals, 3 per cent for navels, as well as breasts, and 2 per cent for legs, buttocks, mouth, lips and teeth.

Body hair, nails, noses, ears, neck and body odour all scored less than 1 per cent.

In a separate analysis, the team looked at sexual preferences for objects associated with the body. Objects associated with the legs and feet had a combined score of 64 per cent, compared with 12 per cent for underwear, and 9 per cent for coats.

The lowest scores went to stethoscopes, wristwatches, bracelets, nappies, hearing aids, catheters and pacemakers.

"These findings provide the first large database in an area where the knowledge is particularly scarce," say the researchers.

Just why and how people develop fetishes is not known, and the researchers say that although many theories have been put forward, none has been fully convincing.

One theory is that there may be a genetic predisposition, but the authors say, "It is unlikely that a particular genetic makeup should result in a preference for specific stimuli such as, for instance, coats, balloons, eyeglasses or headphones – all of which we found in our data."

Early life events may or may not be involved, and Freud had a theory too. "Freud noticed the frequent interest in feet and ascribed this to the notion that feet are a penis symbol."

Other researchers have put forward different suggestions. One suggested that the feet and the genitals occupied the same region of the brain and might therefore have some overlap.

It has also been suggested that foot fetishism increases when there are epidemics of sexually transmitted diseases, presumably as a safety exercise. Yet another theory is that it also starts in childhood, and that playing on the floor around the mother's feet and shoes is the trigger.

Time watchers

WOMEN ARE GETTING FASTER, AND closing the gap on men.

Time-measuring researchers have discovered that the gap between men and women on reaction times, a measure of speed of thinking, has halved over the last 50 years.

The gap in reaction times – also seen by some as a marker for intelligence – is now down to 20 milliseconds, and researchers predict that women will become faster than men within 25 years.

The research also shows that women are closing the physical speed gap on men. In swimming, the difference between the men's and women's records has dropped from 12.41 per cent in 1936 to 9.27 per cent in 1976, and the difference in some events is now down to 5.2 per cent.

Researchers say that emancipation of women has led to the changes, with more women driving, working and taking part in elite sports.

"We think it is all about practice. Reaction times improve with practice, and women have increasingly engaged in activities that provide the opportunity to practice reaction times," says Professor Irwin Silverman, of Bowling Green State University, who led the study reported in the journal *Sex Roles*.

In the research, reaction times between men and women over the last 75 years were compared using data from around thirty individual investigations worldwide. Most of these investigations involved measuring reaction times in laboratory experiments where volunteers reacted to visual or sound stimuli. The researchers also analysed the difference in speed of male and female athletes over 75 years.

The results show that the later the measurements were made, the smaller the gap between men and women. Similar reductions were found for physical performance too. In the Olympics 100 metres, for example, the difference in the winning speeds for the finals for men and women has dropped by 64 per cent since 1928.

"Around 50 years ago it was showing 40 milliseconds, now it is down to 20 and getting smaller. Some research is showing that in some age groups women are already faster. One new study coming out soon shows the gap is below 20 milliseconds," says Professor Silverman. "Projections suggest the gap would disappear in about 25–30 years. Women after all have a natural advantage over men in reaction times because women are on average smaller than men; thus, the neural impulses involved in the production of a motor response have less far to travel in women than in men.

"Changes are taking place all around us which have helped women to cut reaction times. Over the period of time we looked at, women's participation in society has increased significantly. There are, for example, as many women drivers as men, and the number of women in athletics has increased significantly. All those kinds of things have a cumulative effect which has helped increase reaction times."

Another possibility of course is that men are getting relatively slower: "In children the rate of increase in obesity has been greater for boys than girls. Therefore, for motor tasks there should have been a slowing in speed in both sexes, but the rate of slowing should have been greater in boys than girls," says Professor Silverman.

Reaction times may also be connected with speed of mental processing, and IQ: "That is controversial, but it has been suggested because reaction times are about speed of processing," he says.

Making men fat

IF YOU WANT AN INTELLIGENT, trustworthy husband who will be faithful and look after the kids, think big.

Oversize men are seen as brighter, and more reliable and friendly, than thin men, according to new research. And they are also rated as making better parents and mates.

When researchers used computer technology to add 100 pounds to the image of a man, they found that both men and women rated the heavier man as having a number of advantages over the thinner man.

One theory is that thin men are seen as attractive, and that arouses suspicions about trustworthiness and the greater likelihood of them walking out on relationships.

In the research reported in the Oxford-based journal *Personality and Individual Differences*, the psychologists carried out an experiment in which they presented two images of the same man to men and women volunteers. In one of the images, the man appeared to be 7 stone heavier.

The volunteers were then asked to rate each of the men for 25 different personality traits, and for life success, parenting skills and other factors.

When the researchers analysed the results, they found clear differences between ratings for the thin and the larger man.

The thin man was rated as more attractive and enthusiastic, likely to have greater career success that the larger man, and to be more socially desirable.

But the heavy man was considered more intelligent, friendly and trustworthy. He was also rated as having better parenting skills, and greater mate potential.

Just why the heavier man should attract higher ratings in these areas, when the popular image of the overweight person is of someone who is lazy, greedy and selfish, is not clear, but the researchers from Bucknell University in America put forward a number of theories.

"The thinner man may be perceived as less trustworthy due to the relationship between beauty and infidelity. Attractive individuals have more opportunities for sex or better sex, are perceived as more likely to cheat on their mates, and are more likely to desert their current partners, which would not be a characteristic of a good parent or a good mate," say the researchers.

They say overweight men may be seen as more intelligent because they are regarded as having fewer opportunities for socializing: "Individuals may feel an overweight man by virtue of being unattractive would not have the social opportunities that a thin man would have and consequently would devote more time to honing his intellect," they say.

As for the greater friendliness attributed to the larger man, the only explanation here is that it's down to the stereotype of all fat people being jolly.

Telling jokes

MEN REALLY DON'T LIKE FUNNY WOMEN. When they say they like a woman with a sense of humour, what they really mean is someone who laughs at the man's jokes.

The researchers say their findings explain why men claim they value a partner with a good sense of humour, yet do not rate funny women as more desirable. And when a woman laughs at a man's joke, it is a sexual signal of attraction, and the greater the laughter, the bigger the interest.

"Our finding suggests that men prefer sexual partners who appreciate their own humour because that response signals sexual interest," say researchers who carried out the study reported in the journal *Evolution and Human Behavior*.

"If you ask men and women what they look for in a mate they both say a good sense of humour. But when you explore what they mean by a good sense of humour, as we have done, you find men mean they want someone who is going to laugh at their jokes, whereas women tend to mean they want someone who makes them laugh."

The researchers say the findings help to explain why men have been shown not to appreciate women comedians.

New research by psychologists from universities in Canada and America investigated gender differences in men and women in their early twenties.

The volunteers were presented with different scenarios and put through a battery of tests with the idea of teasing out any specific gender differences. The researchers looked at men's and women's responses to humour production and receptivity. Preferences were measured, and the team also looked at whether the preferences change in different types of relationships – long lasting, one-night stands, dates and short-term relationships.

The results show that only women valued their partner's ability to produce humour. They show too that women valued a partner's production of humour as much as they valued a partner's receptivity to their own humour. Men, on the other hand, valued a partner's receptivity to their own humour more than a partner's production of humour.

"Comparing preferences between the sexes for each of the sexual relationships, women showed a stronger preference for humour producers than did men for dates and long-term relationships but not for short-term relationships, one-night stands and friendships," say the researchers.

"Our experiments suggest that the sexes differ in the value they place on a partner's humour production and receptivity. The results suggest that women value a partner who can produce humour and who is receptive to their own humour, whereas men value only a partner's receptivity to their own humour.

"Therefore, the apparent discrepancy between men's valuation of a good sense of humour and their lack of attraction to funny women appears to be resolved: Men's representation of partners who have a good sense of humour does not encompass their ability to produce humour."

They add, "Furthermore, when forced to choose between humour production and humour appreciation in potential partners, women valued humour production, whereas men valued receptivity to their humour."

The researchers say that humour is linked to sexual selection: "We have provided evidence that sexual selection may have influenced humour production because it is specifically preferred by women in relationship partners. Men's reported preferences for humorous partners may be the result of sexual selection shaping a male preference for partners who signal sexual interest through humour appreciation."

Dr Martin of the University of Western Ontario says it may be that a man's humour skills are a sign of good genes.

"From an evolutionary point of view, the theory would be that the ability to create humour might have been an indicator of intellectual ability and creativity, good genes in males," he says. "Females are the ones who are choosy so males have to impress their potential mates, so a good sense of humour would be seen as an indicator of a number of qualities that the female would be looking for. Men are not as choosy. They are willing to take whatever they can get. Looking for someone who can make you laugh is more demanding than just finding someone who will laugh at your humour.

"Women are therefore setting a higher standard. They are saying to the man that he has to create humour, which is a lot harder than laughing at someone else's humour. Someone with a good sense of humour is easy to get on with.

"One of the reasons why men don't like female comedians may be that humour is seen as a masculine thing. When men joke they are competitive, whereas women are more anecdotal. Men see being funny as a male thing," says Dr Martin.

Other research at Freiburg University in Germany supports the idea that men and women tell different kinds of jokes.

It confirms that women's humour is anecdotal and that female jokes invite people to laugh with them, not at them: "Women joked about shared experiences of disappointment, of having to deal with difficult people, and of overcoming the constraints in their lives. A closer look at funny stories about personal misfortunes reveals that in the context of female friendships, the story-teller does not invite the listener to laugh 'at' her, but rather to laugh 'with' her about some absurd aspect of life.

"If everyone jokes about their own faults, ties will grow stronger among the jokers. Further, absurd humour threatens no one. Research shows that women enjoyed the absurd potentials of jokes more than they did the aggressive potentials."

Size does matter (all 5.3 inches)

ALMOST HALF OF MEN WISH their manhood was bigger, and seven out of ten rate their own as average, according to the world's first survey of sex, size and satisfaction.

Researchers, who calculate the average length when in action to be 5.3 ins, or 13.5 cm, also found that men who were satisfied with their size were more likely to be taller, thinner, and more confident and happier about their looks and appearance.

Results also show that 14 per cent of women want their partners to be bigger, while only 2 per cent want a smaller model.

"This is the first large-scale study of the associations among men's penis size, satisfaction, and body image," say the researchers, whose results appear in the medical journal *Psychology of Men & Masculinity*.

The international study, based on a 27-question Internet survey completed by more than fifty thousand men and women aged 18 to 65, found that most men – 66 per cent – rated their size as average, 22 per cent as large, and 12 per cent as small.

"For many men, being average wasn't good enough. Among men who rated their penis size as average, 46 per cent wanted to be larger," say the researchers from the University of California. "We found that many men – 45 per cent – desired a larger penis."

The results also show that seven out of ten women rated their partner's size as average: "Some women viewed their partner as large (27 per cent), and few women perceived their partner as small (six per cent)."

But despite men's worries, the results show that most women – 84 per cent – are satisfied with their partner's size, and only 14 per cent wanted their partner to be larger.

The researchers also found that men with larger-than-average penises were more confident. They report no evidence to support folk beliefs that size is related to foot size and hand size, although taller men were larger, and fatter men smaller.

On actual size, the researchers say that the average operational length was 5.3 ins (13.5 cm), with 68 per cent of men measuring between 4.6 and 6 ins. A total of 13.5 per cent were between 3.8 and 4.5 ins (9.7 cm and 11.4 cm), and 13.5 per cent between 6.1 and 6.8 ins (15.5 cm and 17.3 cm). Only 2.5 per cent of men were over 6.9 ins (17.5 cm) long, and 2.5 per cent were under 3.7 ins (9.4 cm) long.

Studying one-night stands

WOMEN ARE MORE LIKELY THAN MEN to have regrets about one-night stands.

They are also less likely than men to want their friends to know about what they have done, and are more worried about loss of reputation, according to research based on the biggest academic study of one-night stands.

One theory as to why women have greater regrets after a one-night stand is that their pride may be dented by the man not being appreciative enough of their sexual generosity.

"Our results suggest that women find the experience of casual sex less satisfactory than men. This may explain why women are more reluctant than men to engage in short-term relationships," says psychologist Professor Anne Campbell of Durham University, who led the study.

In the study, the psychologists looked at the attitudes of those who had had a one-night stand, defined as a sexual relationship that progressed no farther than copulation. The study examined men's and women's responses to one-night stands that they have actually experienced rather than the hypothetical situations looked at by other researchers.

The research was based on a sample of 1,909 men and 1,454 women of whom 2,956 or 88 per cent were heterosexual. Of these heterosexuals, 1,743 or 59 per cent had experienced a one-night stand – 998 men and 745 women. Of these, 265 men and 134 women were in steady relationships at the time of the one-night stand. Some 42 per cent were aged 17–25, 40 per cent were 26–40, and 13 per cent were 41–60.

As part of the study, the men and women were asked to rate aspects of their morning-after feelings.

Results of an analysis of 233 men and women who gave more detailed comments show that 59 per cent of men and 28 per cent of women had positive recollections of their one-night stand. Descriptions made to researchers included: euphoric, exhilarating, fun, excitement and adventure. Twenty-three per cent of men and 58 per cent of women in the sample had some regrets and said they would not repeat the experience.

More women than men regretted the fact that they had let themselves down, and were worried about their loss of reputation. They were also more likely to feel they had been used.

More men than women secretly hoped their friends would hear about the escapade, were more sexually satisfied, and felt more confident about themselves after the one-night stand. Feelings of being flattered and successful were about the same in both sexes.

"Women rate the experience of one-night stands both less positively and more negatively than do men," says Professor Campbell. "A feeling, for women, of having been used was the strongest sex difference, followed by a sense of having let oneself down and loss of reputation."

The researchers say the results suggest that women's motivation for one-night stands is not an attempt to test out or secure a long-term or back-up partner.

But if most women have a barely positive view of one-night stands, why do they indulge in them?

"Regardless of how the women felt after the experience, the fact remains that at the time they chose to have a transitory sexual encounter," says Professor Campbell.

Ear testing

SWEET NOTHINGS SHOULD BE WHISPERED into the left ear for maximum effect.

Emotional words are processed better by the left ear than the right, according to researchers whose work shows that men and women were able to accurately identify and recall more than 70 per cent of ten emotional words, such as "love", "kiss" and "passion", perceived by their left ear, compared with 58 per cent for those perceived by the right.

They were also able to recall more negative words such as "sad" and "angry" when these were spoken into the left ear, which is controlled by the right side of the brain, the so-called emotional side.

The findings may help to explain why mothers mostly cradle babies on their left side, closer to the left ear. According to researchers at the Hammersmith Hospital, "Left-cradling not only directs maternal communication to the infant's right hemisphere but also facilitates affective feedback to the maternal right brain." They may also account for why some research shows that listening to music with the left ear can be more stimulating.

In the research, carried out by academics at Sam Houston State University in America, 100 men and women wore earphones to listen to a number of emotional and neutral words being read in each ear separately. The words were read without any emotion.

Some time later, the volunteers were asked to write down the words they had heard in each ear. The results show that the recall accuracy for emotional words was 70.43 per cent for the left ear, and 58.67 per cent for the right ear.

They also show that non-emotional words, including "lemonade", "machine" and "engine", were recalled more accurately when picked up by the right ear.

"This study examines the detection and classification of emotion words – angry, sad, happy or positive and negative emotion words – that are articulated in neutral tones. We found a left ear advantage and the results are consistent with the idea that the right hemisphere of the brain is more adept at this task," say the researchers.

"It also provides strong support for the idea of a stronger memory for emotion words presented to the left ear. The findings are consistent with the role of the right hemisphere of the brain in the perceptions of emotional information."

Other research shows that music is also processed more effectively by the left ear and the right side of the brain, and that melodies are recognized more accurately by the left ear. According to a study at Middle Tennessee State University, the left side of the brain processes information more logically, while the right side is more intuitive.

7

HOW TO BOIL AN EGG, USE DANDRUFF, AND WATCH CABBAGES GROW

Watching cabbages grow

CABBAGES ARE AN IMPORTANT CROP. More than 50 million tonnes of the green leaf vegetables are harvested each year worldwide, and they are a major source of dietary nutrients.

But it is also a crop that is very vulnerable to pests and other problems that can affect its growth, and hence overall productivity and profits – production in America alone is estimated to be worth $500 million.

Cabbage can be infested by insects as diverse as aphids, leafminers, thrips, whitefly, cabbage maggots, beetles, true bugs, caterpillars, cabbage loopers and diamondback moths, as well as slugs and snails.

While pesticides and other tactics can be used to tackle these threats, there have been problems in monitoring how well the cabbages are growing on large commercial plantations. In China, for example, which is the world's biggest producer, it's estimated that cabbage plantations cover more than one million acres, so some kind of real-time monitoring could be vital.

And that's why researchers have been experimenting with aerial photography to watch cabbages grow and determine how they respond to various pesticides and treatments.

In the work at Texas A&M University, researchers first planted cabbages and then treated them with fifteen different insecticides. Each insecticide was applied four times on 81 individual experimental plots seeded with *Brassica oleracea var. capitata L.*

After the final spraying, and with the cabbages growing well, the researchers took to the air in a Cessna 404 aircraft with a camera port in the floor that was used to take aerial photographs at an altitude of 460 metres above ground level using green-sensitive film.

The resulting digital images were then used to measure plant diameter, head diameter, plant weight and head weight. The researchers found they were easily able to distinguish between a healthy plant, which was 44 cm in diameter, and stressed cabbage, at 25 cm.

They were also able to assess the performances of the individual insecticides used.

"Compared with traditional ground observation and measurement approaches, aerial photography is more effective and efficient if a large number of plots or treatments are to be evaluated over large fields," say the researchers.

"This is one of the first evaluations of remote sensing techniques for estimating cabbage plant growth parameters and the results were encouraging. More experiments are needed to evaluate this technology for mapping cabbage growth and yield variations and for assessing the efficacies of different insecticide treatments for controlling cabbage insects."

Meanwhile, researchers at the University of Newcastle have found that garlic could be used to tackle the number-one enemy of cabbages, slugs.

Garlic, used by the ancient Egyptians to keep vampires at bay, has been used in "companion planting" strategies for hundreds of years. Monks used to site garlic next to their vegetable crops to keep unwanted pests away. Garlic has been shown to be one of the most effective slug killers, and the Newcastle researchers suspect it may have an adverse effect on the creatures' nervous systems. Their work found that garlic kills slug eggs laid in the soil.

Slugs and also snails cause millions of pounds' worth of damage as they munch their way through food crops and plants, particularly those in cool, temperate climates like those of the UK, northern Europe and north-west America. Even more millions of pounds are spent trying to control them – the estimated overall cost to the UK is around £30 million.

The Newcastle University scientists looked at how applying a liquid containing garlic extract to soil affected slugs' and snails' movement through it. They also measured damage to a Chinese cabbage leaf and found that garlic largely prevented the leaf from being eaten and killed a very high number of the creatures.

"We need to carry out more tests to find out its commercial potential. We want to find out how garlic affects other creatures living in the soil, the right concentration to use, how it affects the taste of food once it has been used on crops, and many other things," say the researchers.

Egg-boiling

BOILING AN EGG IS A TASK that you may well think requires little science and research.

Three minutes gets a runny yolk suitable for toast soldiers, and ten minutes delivers a hard-centred egg for salads: simple.

Not so, according to Dr Charles Williams, a physicist at Exeter University. It is slightly more complicated. In fact, he has been using a formula for calculating the time required to boil an egg given its size, weight, density and initial temperature, and, of course, taking into account the specific heat capacity and thermal conductivity of the egg.

In his calculations, Dr Williams investigated a number of factors, including the make-up of the shell, white and yolk of the egg.

The shell, it seems, accounts for about 9 to 12 per cent of the total egg and it has 7,000–17,000 tiny pores distributed over its surface. The white or albumen accounts for most of an egg's liquid weight, about 67 per cent. It consists of four opalescent layers of alternately thick and thin consistencies. The white of a freshly laid egg has a pH between 7.6 and 7.9 and an opalescent or cloudy appearance owing to the presence of carbon dioxide.

The yolk makes up about 33 per cent of the liquid weight of the egg. It contains all of the fat and all the vitamins, and slightly less than half of the protein.

Dr Williams went to great lengths to arrive at a formula for soft-boiled eggs: "To obtain a simple formula the problem must be idealized, so the egg will be treated as a spherical homogeneous object of mass M and initial temperature T_{egg}. If the egg is placed straight into a pan of boiling water at T_{water}, it will be ready when the temperature at the boundary of the yolk has risen to $T_{yolk} \sim 63°C$. With these assumptions, the cooking time can be deduced by solving a heat diffusion equation," he says.

According to this formula, a medium egg (weighing around 57 grams), which has been taken straight from the fridge (at 4°C), takes four and a half minutes to cook, but the same egg would take three and a half minutes if it had been stored at room temperature (21°C). If all the eggs are stored in the fridge, then a small (size 6 or 47 gram) egg will require four minutes to cook, and a large egg (67 grams) five minutes.

His research shows that even these times can be made more accurate. For a more accurate time, he says, the different thermal properties of the white, yolk and shell all need to be taken into account.

"One would need to treat the egg as three concentric ellipsoids with Dirichlet boundary-conditions at the water–shell interface and Neuman boundary-conditions for the shell–white and white–yolk parts. Changes in its thermal properties when the white changes state from solid to gel, and the latent heat associated with this change, would also need to be accounted for," he says.

But the answer to even that formula may be inaccurate on some occasions. His work also shows that it takes noticeably longer to boil an egg on a mountain than it does at sea level. That's probably because the boiling point of water falls with decreasing atmospheric pressure.

Rules for hard-boiling an egg are, of course, completely different.

The first, he says, is to place the egg into a pan of boiling water and cook for twice the time required to soft-boil, then plunge it into cold water. Alternatively, he suggests, put the eggs into a pan of cold water, bring to the boil, then remove the heat and let the pan stand with its lid on for about seventeen minutes, and then cool.

Both methods are intended to prevent the yolk temperature getting too high (about 70°C), when hydrogen sulphide generated by the decomposition of sulphur-containing amino acids in the white will react with iron in the yolk, causing a grey-green film of ferrous sulphide to form on the surface of the yolk.

Sadly, none of the methods researched by Dr Williams appears to satisfy his palate: "My preferred method of boiling eggs is to steam them. Use about one cm depth of boiling water in a pan with a close-fitting lid. The eggs rest on the bottom of the pan which reduces bumping and presumably reduces the chance of cracking. Since only a small amount of water needs to be boiled the method also saves time and energy."

The dandruff detectives

DANDRUFF COULD BE THE DOWNFALL of bank robbers.

Research based on analysing the contents of dirty T-shirts and tights has given the FBI a new way to tackle crime – dead skin cells.

"Skin cells obtained from scraping an item of clothing could contain sufficient DNA to potentially determine the source of the wearer," say the FBI researchers. "In a bank robbery, for example, when only masks or gloves are recovered, it may be the only biological material available for forensic DNA analysis."

Because skin cells are constantly shed, it is likely that the debris in clothing collected from suspects also contains skin cells from the individual who wore the garment. These cells contain nuclear DNA and may have value as evidence. Even tiny amounts of biological material may yield sufficient quantities of DNA for analysis, say the researchers. Just 1 milligram of dandruff may be enough.

Hairs and fibres and other trace elements from suspects and victims are routinely collected at crime scenes for DNA examination, but testing may miss material like dandruff and skin cells, which could also be used to identify the wearer.

To test the theory that cells could be collected and used, FBI researchers in Washington have been testing T-shirts and hosiery which are worn by laboratory personnel for a period of time and then scraped for trace evidence.

The employees wore a freshly laundered item of clothing for one day. Women wore hosiery, and males wore T-shirts. After the workday, the items were collected and stored in clean paper or plastic bags for analysis.

All items were processed for trace evidence by scraping the inside and outside, and DNA analysis was carried out. The people taking part in the study, their cohabitants and the people carrying out the testing were also tested. Cells were collected from each item using a sterile swab moistened with water. The T-shirts were swabbed around the neckline.

Results show that enough material could be harvested from the clothing for DNA analysis that identified the wearer. "This study demonstrates that trace evidence debris can provide a sufficient quantity and quality of DNA to potentially identify the wearer of an item," say the researchers.

So accurate was the testing that it was possible to identify people that the wearer had come into contact with. In some cases the researchers could identify the persons, usually a partner; in others they could not.

In other cases it was not possible to explain how the other DNA came to be on the clothing and further work is under way. "Additional studies are being conducted to test the possibility of DNA carryover in washing machines during the laundering process in an effort to explain the presence of cohabitants' DNA on items of clothing," says the researchers.

Academics at the University of Granada in Spain have been working along the same lines. They looked at whether or not DNA could be extracted from dandruff, and how. Results show that sufficient quantities of DNA – more than 30 to 40 nanograms – could be obtained from as little as 1.0 to 1.5 milligrams of dandruff. "Dandruff can be considered a potential source of DNA for forensic identification," they say.

Egg-cracking

SCIENTISTS HAVE FINALLY CRACKED IT.
A team of researchers have worked out the exact force required to break an egg, and which are its strongest and weakest parts.

According to the academics, the rupture force of a hen egg depends on various properties, including specific gravity, mass, volume, surface area, thickness and shell weight: "Eggshells must be strong enough to prevent cracking, weak enough to allow the chick to break through when hatching and thin enough to allow gas exchange," they say.

Although egg-testing is popular in school science experiments, largely because of its destructiveness and the messy nature of the job, the researchers say there is a lack of published scientific research on how eggs behave under pressure.

In the research, the academics used 270 eggs and made a series of measurements, including pressing them between steel plates and then calculating the compression speeds needed to break the shell.

Resistance of hen eggs to damage was worked out by measuring the average rupture force, energy and firmness of different parts of the eggs.

Results show that the greatest amount of force required to break the eggs was required when they were loaded from top to bottom rather than side to side. Average shell thickness was 0.344 to 0.351 mm.

To test shell breaking strength, the eggs were placed between two plates and put under increasing pressure until the shell broke. The force necessary to break the shell was found to be around 30 newtons.

The researchers from the University of Gaziosmanpasa in Turkey say the results are important because of the large number of eggs packaged and transported and thus vulnerable to breaking.

"The mechanical properties are necessary considerations in the design and effective use of the equipment used in the transportation, processing, packaging and storage," they say.

Eggshells are about 94 per cent calcium carbonate with small amounts of magnesium carbonate, calcium phosphate and other organic matter, including protein.

Shell strength is influenced by two factors. First, the hen's diet, particularly its calcium, phosphorus, manganese and vitamin D intake.

Second, egg size, which increases as the hen ages, while the mass of shell material that covers it stays fixed. Hence the shell is thinner on larger eggs.

Up to 17,000 pores are distributed over the shell surface. As the egg ages, moisture and carbon dioxide diffuse out and air diffuses in through these, causing the air cell to grow and the net mass to decrease. The shell is covered with a protective coating called the cuticle. By blocking the pores, the cuticle helps to preserve freshness and prevent microbial contamination of the contents.

Pricking the blunt end of an egg does not, as is commonly thought, prevent the shell from cracking, say the researchers; it relieves pressure and reduces the extent to which the egg is forced out through any crack that does develop. Adding vinegar and/or salt to the water also helps by causing the emerging liquid to coagulate, thereby plugging the gap more rapidly than simple hot water would do.

Why wallpaper never comes off in one piece

THERE YOU ARE, DESPERATELY TRYING to finish stripping the wallpaper.

You spot a large loose piece hanging invitingly high up on the wall. It seems to be wanting you to grab on it and pull. So you do, hoping as always that the paper will obligingly peel off all the way to the ground.

But it doesn't. As you pull it down the wall, it gets narrower and narrower, and you're left holding a perfectly formed, but frustratingly small, triangular piece of paper.

No matter how many times you try, and however great the initial optimism, the wallpaper always peels off in the shape of a triangle long before it reaches floor level.

"You want to redecorate your bedroom, so you yank down the wallpaper. You wish that the flap would tear all the way down to the floor, but it comes together in a triangle and you have to start all over again," says Pedro Reis, a Massachusetts Institute of Technology researcher who has made a study of wallpaper behaviour.

Ever since the Chinese began gluing rice paper on to their walls as long ago as 200 BC, wallpaper strippers have been confounded and frustrated by their inability to get the paper off in large sections.

Wallpaper technology may have advanced over the centuries, incorporating compounds that make it washable, pre-pasted and long lasting, and it may even incorporate digital images, but when it comes to stripping, it still never comes off in anything but triangular pieces.

But why? According to research by Reis and his wallpaper-stripping colleagues from the Centre National de la Recherche Scientifique (CNRS) in Paris, and the Universidad de Santiago, Chile, it's about the wallpaper rip phenomenon or WRIP, and it is all down to basic physics.

It happens when two cracks, such as the break lines in a piece of wallpaper, move inevitably towards each other and meet at a point. It also applies to adhesive tapes and plastic sheets, and old posters. Investigations by the researchers show that the skins of tomatoes and grapes tend to form triangles when peeled off too.

After much laboratory work, the researchers found that the triangular tears are the unstoppable consequence of interactions between three properties of adhesive materials like wallpaper – the elasticity or stiffness of the material, how strongly the adhesive sticks to a surface, and how tough it is to rip.

The bottom line is that as the strip is pulled, energy builds up in the fold that forms where the tape is peeling from the surface. The tape can release that energy in two ways – by peeling off the surface it is attached to and by becoming narrower. According to the research, it does both, resulting in WRIP.

As the researchers point out, wallpaper, torn posters and tomato skins may seem like strange research areas for physicists and applied mathematicians, but their work may well have important real-life uses.

"We can really learn things that will be useful for industry and help us understand the everyday world around us," said Reis. "This shape is really robust, so there must be something fundamental going on that gives rise to these shapes."

The researchers were able to develop a formula that predicts the angle of the triangle formed, based on those three properties. One possible industrial application is that engineers could use this formula to calculate one of the three key properties, if the other two are known.

Reis and his collaborators got the idea for the project after noticing consistent tearing patterns in plastic sheets, such as the plastic wrapping of CDs.

The same kind of triangular shapes can also be seen in the work of French artist Jacques Villiegle. His art mainly consists of posters taken from the streets of Paris and other French cities, complete with the same sort of rips that the researchers studied.

Watching goalkeepers dive

EVER WONDERED WHY GOALKEEPERS DIVE at penalty kicks when they have no chance of stopping the ball?

According to an academic study, it's because it makes them feel better when the ball hits the back of the net. They have worked out that 94 per cent of goalies dive, either to the left or the right.

But diving to left or right doesn't work, it seems, and the academics have calculated that the best strategy is for the 'keeper to stay in the centre of the goal.

"A goal scored yields worse feelings for the goalkeeper following inaction – staying in the centre – than following action – jumping – leading to a bias for action," say the researchers, who analysed 286 penalty kicks in top leagues and championships worldwide.

"We show that given the probability distribution of kick direction, the optimal strategy for goalkeepers is to stay in the goal's centre. Goalkeepers, however, almost always jump right or left."

The researchers, who say it takes the ball only around 0.2 seconds to travel the 11 metres from the penalty spot to the goal, say the 'keeper cannot afford to wait until he sees clearly in which direction the ball is kicked before choosing what to do.

"He has to decide whether to jump to one of the sides or to stay in the centre at about the same time that the kicker chooses where to direct the kick," says the report.

To work out the best strategy, with the highest probability of stopping a goal when the ball is sent in that direction, the researchers took the number of balls stopped when the goalkeeper chose each direction and divided it by the total number of dives in that direction.

The researchers say the goalkeeper's optimal decision is to choose the direction where his probability of stopping the ball is maximum.

Goalkeepers dived to the left 141 times and stopped 20 balls when doing so, giving a 14.2 per cent stopping rate. Those diving to the right had a stopping rate of 12.6 per cent. Those in the centre accounted for six stopped kicks and let in 12 goals, a stopping rate of 33 per cent.

"While it is optimal for goalkeepers to choose to stay in the centre, however, they almost always choose to jump to one of the sides! Only in 6.3 per cent of the cases do they stay in the centre. This suggests that goalkeepers might exhibit biased decision making," say the researchers from Ben-Gurion University and the Hebrew University, Israel.

They add, "We propose that the reason for goalkeepers not staying in the centre is action bias.

"A goal being scored is perceived to be worse when it follows inaction rather than action. If the goalkeeper jumps and a goal is scored, he might feel 'I did my best to stop the ball, by jumping, as almost everyone does. I was simply unlucky that the ball headed in another direction or could not be stopped for another reason'.

"On the other hand, if the goalkeeper stays in the centre and a goal is scored, it looks as if he did not do anything to stop the ball when the norm is to do something, to jump. Because the negative feeling of the goalkeeper following a goal being scored is amplified when staying in the centre, the goalkeeper prefers to jump."

The researchers also add a note of caution: "We want to stress that the argument that the goalkeepers are better off choosing to stay in the centre is based on the current distribution of kicks. If goalkeepers will always choose to stay in the centre, kickers will start aiming all balls to the sides, and it will no longer be optimal for the goalkeeper to stay in the centre."

Why fat people are looked down on

OBESE PEOPLE ARE STIGMATIZED BECAUSE being too fat is mistaken by the observer's brain as a sign of disease.

Our immune system, it seems, can be triggered into action at the sign of obesity because it doesn't like the look of what it sees, and associates it with infection. Just as it orchestrates attacks on viruses and bacteria, zaps foreign bodies and triggers nausea at the hint of bad food, it sends out signals of disgust in some people at the sight of an obese body.

"Antipathy toward obese people is a powerful and pervasive prejudice in many contemporary populations. Our results reveal, for the first time, that this prejudice may be rooted in multiple, independent mechanisms. They provide the first evidence that obesity serves as a cue for pathogen infection," say University of British Columbia researchers.

They say that a behavioural immune system appears to have evolved which is designed to detect body signs that are related to disease, such as rashes and lesions. The sight of them triggers disgust as well as negative attitudes and avoidance.

The system errs in favour of overreaction because not reacting to a real danger could be fatal. For that reason, there is a bias towards inferring that healthy people are diseased, rather than the reverse. That's why, it's suggested, many people have negative views about facial birthmarks.

But could such an overreaction also explain the stigmatization of obesity?

To test the idea, the researchers carried out a number of experiments, including word associations, and tests where they compared the reactions and views of men and women to obesity. One group was more concerned about their health and considered themselves more vulnerable to ill health than the other.

The results show that people who agreed with comments like "It really bothers me when people sneeze without covering their mouths" were more likely to agree with statements like "If I were an employer looking to hire, I might avoid hiring a fat person".

"These results support the hypothesis that concern with pathogen transmission predicts antipathy toward obese people. The correlation was especially strong following visual exposure to obese individuals," says the research report.

"Results revealed that the perception of obesity inspires greater antipathy when perceivers feel more vulnerable to disease. There is no readily apparent alternative explanation that can account for the full set of results found across both studies.

"Obese people were implicitly associated with disease-relevant concepts," say the researchers. "These results offer the first experimental evidence indicating a causal impact of pathogen concern on perceptions of obese people. Moreover, they further indicate that pathogen-avoidance processes are psychologically independent of other processes that have been previously linked to antifat attitudes."

They say that the findings could be useful for tackling prejudice: "Antipathy toward obese people is a powerful and pervasive prejudice, and our results reveal, for the first time, that this prejudice may be rooted in multiple, psychologically independent mechanisms. This is sobering but encouraging because it provides clues toward the possible reduction of this form of prejudice."

They add, "Previous research shows that this prejudice may be reduced through interventions that focus on attribution processes and suggests that it might also be ameliorated through interventions that focus on individuals' often-irrational concerns about infectious disease."

Bad driving runs in families

RESEARCH AMONG MARRIED COUPLES SHOWS that "for better, for worse" applies to driving behaviour too.

Married couples drive like each other, say the researchers, who found that women married to men who were aggressive and reckless were more likely to show the same kind of behaviour.

"The higher the level of recklessness and aggressiveness of men, the higher level of such driving styles can be expected for their women partners. Their partner's driving style may contribute to women's tendency for reckless driving behaviour and the likelihood of their committing traffic violations," says Dr Orit Taubman.

Just why is not clear, but researchers say there are two theories. Couples may grow to drive in the same way simply as a result of copying their partner's behaviour over time.

Or it may be that a shared driving style is one of the characteristics that attracted them to each other in the first place.

"Driving style might represent behaviour in which individuals observe specific overt behaviours of their partner, and may either adopt part of that behaviour, or perhaps even choose their partner because of it," says Dr Taubman, who led the study at Israel's Bar-Ilan University.

In the research, reported in the academic journal *Transportation Research*, the team investigated the driving styles of more than eight hundred married men and women aged 19 to 68, and compared the data on husbands and wives. The team looked at a variety of factors, including choice of driving speed, and levels of attentiveness and assertiveness.

Describing this as the first research into the driving styles of men and women who are intimate partners, its authors say it opens up the way to safe driving programmes aimed at couples and families.

Watching the weather

ALMOST EVERYTHING THAT CAN HAVE an effect on education has been investigated down the years, from family income and genes, to personality and diet.

But until now, the weather, and in particular clouds, have been unexplored territory for academics.

But not any more, for researchers have discovered that cloudy days may affect the chances of being accepted for university.

According to research based on cloud counting, those who make admission decisions are unknowingly influenced by the state of the weather on the day they consider applications.

Using what they call a nerd index, researchers have discovered that a change in cloud cover can increase or decrease a candidate's chances of admission by 11.9 per cent.

They found that more weight was given to academic achievements on cloudy days, and greater weight to non-academic, social and sporting activities on sunny days.

"It shows that professional university admission reviewers weight the attributes of applicants differently, depending on how cloudy the day is when they happen to be reviewing them," says Dr Uri Simonton.

In the research he analysed hundreds of university admission decisions, looked at the records of applications, and compared these with the weather on the days the decisions were made. He also worked out what he calls a nerd index, a figure that results from dividing the academic score of the students by the social score.

"The nerd index was significantly higher for students admitted on cloudier days than for those admitted on sunnier ones," says the report.

And the implications of being considered under the wrong weather conditions are significant. "An applicant's predicted probability of being admitted increases by 11.9 per cent if her application is read under optimal versus worst possible cloud cover. Applicants need to increase their academic rating by 28.5 per cent in order to obtain a similar gain in admission probability."

Just why is not clear, but one theory is that emotions are influenced by the weather. People are sadder when it is cloudy, and sad people are inclined to become more analytical. Other research shows that people tip less on cloudy days.

The researchers say that experts have been shown to make different judgements on the same data on different occasions, but that until now it has been blamed on fatigue, boredom and distraction.

Weighing up climate change

OBESE PEOPLE MAY BE KILLING the planet as well as themselves.

Researchers have calculated that the huge worldwide rise in the number of obese people is increasing greenhouse gas production.

They have worked out that the average 5 kg rise in weight per person that has occurred over the past fifteen years is responsible for 5 to 10 per cent of all greenhouse gases.

The researchers, who suggest the introduction of a fat tax on food to tackle the problem, looked at a range of costs associated with obesity, including increased fuel to transport heavier people around, extra energy use to manufacture more food, increases in sedentary energy-using activities by the obese, such as TV watching, additional meat production, more livestock, greater amounts of fertilizer and manure, and increased waste production.

"It is a conservative estimate because there are a number of other things we did not take into account. We estimate that obesity is responsible for five to 10 per cent of the 20 billion tonnes of greenhouse gases produces each year by industrialized countries," says Dr Axel Michaelowa of Zurich University, who led the study.

"Obesity increases greenhouse gas emissions through higher fuel needs for transporting heavier people, emissions due to additional food production, methane emissions from higher amounts of organic waste, and other factors. Our aim was to assess the impacts of obesity increase on greenhouse gas emissions."

Latest figures for the UK show that around 24 per cent of the population are obese, and that the number has been increasing for some years; and obesity has been linked to increased risk of cancer, heart disease and other conditions.

In the study, reported in the journal *Ecological Economics*, researchers looked at the impact of a 5 kg or 11 pound average increase in weight. The figure was chosen because that is the average weight increase seen in America and other countries over the last decade and a half.

In terms of transport, carrying heavier people around was calculated to be the equivalent of 10.2 million tonnes of CO_2. The biggest contributions – 8.7 million tonnes – came from larger people being carried in cars. A total of 1.2 million tonnes was down to the added weight of air travellers, and 280,000 tonnes to rail travel.

The researchers say that the increase in obesity is due to increased food consumption, and to a shift to greater consumption of foods containing saturated fats, such as meat and dairy products. They say that since 1964, consumption of meat has risen by 43 per cent and of milk by 14 per cent in industrialized countries.

The researchers calculated the CO_2 emissions for each of the major food products of animal origin and of wheat and bakery products.

They found that average daily food intake in industrialized countries has increased from 12.34 MJ (megajoules) in the mid-1960s to 14.15 MJ. They estimate that, in terms of food production, emissions associated with obesity-causing foods total around 400 million tonnes of CO_2.

More food means more animals, and the researchers worked out levels of methane produced by additional cows, pigs and chickens and their contribution to greenhouse gases. They also looked at the fuel, wheat, fertilizer and chemicals needed to process food.

Higher food consumption also leads to more human waste. They calculate that the waste generation due to the increase in food consumption from 1990 was around 7.5 million tonnes, which is equivalent to 4.5 million tonnes of carbon dioxide.

The researchers also calculated the energy costs of increased TV watching, which has also been associated with the rise in obesity. They found that one extra hour of TV watching is the equivalent of 25 million tonnes of CO_2. For the UK alone the figure is 1.3 million tonnes.

"Obesity is a societal challenge that is on the rise in almost all countries. We have shown that obesity leads to increasing greenhouse gas emissions. In the transportation sector the effect of an average 5 kg weight increase results in additional 10 million tonnes of CO_2 for the OECD countries," say Dr Michaelowa. "On food production, we found emissions of obesity-causing food including meat and dairy products, increased by almost 400 million tonnes of CO_2. Emissions due to obesity total hundreds of million tonnes."

HOW TO CALCULATE WINNERS, SPOT LIARS, AND STOP CHEWING GUM STICKING TO PAVEMENTS

Calculating winners

$$\text{fi } j = ft + (\text{ fi} - fav) - (gj - gav)$$

THAT'S THE EQUATION THAT COULD DECIDE who will win Wimbledon.

The mathematicians who developed it, and whose previous work on betting on single matches is already used by bookmakers, say the equation could be used by punters and by Wimbledon organizers to predict the length of individual matches and who will go through to the next round

"What we have been able to do is to combine player statistics to predict the outcome of tennis matches. Published tennis statistics can be combined to predict the serving statistics when two given players meet. These are then used to predict further match outcomes, such as the length of the match and the chance of either player winning," says Tristan Barnett, who led the team of mathematicians who reported their work in the *Journal of Management Mathematics*.

With men's tennis dominated by serves, the researchers at Swinburne University in Australia combined statistics about serve and return for each player.

Using publicly available data on the top 200 players, the researchers worked out the chances of winning a point on serve, on first service alone, and on return of serve. That is complex because it has to take into account not only the server's ability, but that of the receiver.

"While we expect a good server to win a higher proportion of serves than average, this proportion would be reduced somewhat if his opponent is a good receiver," says Dr Barnett.

When they fed all the data they had into the calculations, the main equation they came up with was designed to predict percentage serving and receiving success. The results also backed the belief that serve-and-volleyers perform best on grass surfaces.

"In simple terms, we take the percentage of points a player wins on serve as the overall percentage of points won on serve for that tournament, which takes into account court surface, plus the excess by which a player's serving percentage exceeds the average, minus the excess by which the opponent's receiving percentage exceeds the average," says Barnett.

The researchers tested their formula in one match between Roddick and El Aynaou in the 2003 Australian Open, and found that it could accurately predict events, including the length of the match.

"It could be useful in a number of areas. In planning daily draws, tournament organizers also have an interest in predicting the chances of each player advancing in the draw, and the probable length of matches," say the researchers. "Long matches require rescheduling of following matches, and also create scheduling problems for media broadcasters."

Eyebrow testing

ARCHED EYEBROWS ARE OUT, AND low brows in.

High eyebrows, once the most popular, could disappear altogether over the next few years, according to research.

Researchers have shown that the most attractive eyebrows are those that curve lower over the eye, reaching a peak height two-thirds of the way along, or about 9 mm on the ear side of the eye. Plastic surgeons are now being urged to keep eyebrows low when they carry out cosmetic surgery.

"Current concepts of brow lift indications need to be reconsidered. The eyebrows are frequently placed too high, with the eyebrow arch in the middle, frequently leaving the patients with an unnatural astonished expression," say the researchers, who report their findings in the medical journal *Aesthetic Plastic Surgery*.

In rankings of the attractiveness of facial features, eyebrows are consistently rated as one of the most important. For some time, thin, high, arched eyebrows have been popular because they are thought to lower a woman's age.

But the research by plastic surgeons at the University of Regensburg, and another report from Johns Hopkins University, shows that lower brows are the shape of the future.

In the German research, the team showed pictures of women with different-shaped eyebrows to around 350 people aged 12 to 85. One of the three basic types of eyebrow is the arched eyebrow with the maximum height in the middle, while the other two both had their maximum height in the last third of the brow, on the ear side, with one slightly higher than the other.

Results showed that younger men and women rated both lower eyebrow positions in all faces to be more attractive, while older people found the arched type more attractive.

"Young people up to 29 years of age judge arched eyebrows to be unattractive and prefer the lower positioned eyebrow with a maximum in the lateral third. This form has become more prevalent over the past several years and can currently be described as the new ideal," say the researchers.

"The data indicate that the beauty ideal of eyebrow

position is currently undergoing a change because younger people prefer a different eyebrow position than older individuals. Because trends are generally established by young people rather than older individuals, it seems plausible to assume that the social trend currently is toward a lower eyebrow position with a maximum in the lateral third.

"The ideal of the arched eyebrows likely will become less important and may even die out in two or three decades with the generation currently 50 years of age or older."

But even the older woman may opt for lower brows: "Many women consider it important not only to look attractive, but also to be seen as modern. Such women probably instead will strive, regardless of their age, for lower placed eyebrows because this form has obviously come into vogue."

Similar findings have been reported by a team at Johns Hopkins University. They interviewed 100 people and asked them to rank 27 photographs for attractiveness: "Our study confirmed that placement of the eyebrow in a lower position than has been previously thought to be ideal, including a lower eyebrow arch position, may actually result in an eyebrow that is more attractive to the majority of people," they say.

```
        Barnes & Noble Booksellers #2614
              2100 North Snelling Ave
                Roseville, MN 55113
                   651-639-9256

STR:2614 REG:003 TRN:5027  CSHR:Erica H

Boringology
    9781905736157
    (1 @ 15.95) Item Cpn 25% (3.99)
        #500741
    (1 @ 11.96)                          11.96
Magic School Bus and the
    9780590446839
    (1 @ 6.99)                            6.99

Subtotal                                 18.95
Sales Tax (7.125%)                        1.35
TOTAL                                     20.30
VISA                                      20.30
    Card#:  XXXXXXXXXXXXX3100
    Expdate: XX/XX
    Auth:   026416
    Entry Method: Swiped

A MEMBER WOULD HAVE SAVED               2.30


                Thanks for shopping at
                    Barnes & Noble

V101.19                        11/20/2009  05:04PM
```

```
                 CUSTOMER COPY
```

Magazines, newspapers, and used books are not returnable. *Product not carried by Barnes & Noble or Barnes & Noble.com will not be accepted for return.*

Policy on receipt may appear in two sections.

Return Policy

<u>With a sales receipt, a full refund in the original form of payment</u> will be issued from any Barnes & Noble store for returns of new and unread books (except textbooks) and unopened music/DVDs/audio made within (i) 14 days of purchase from a Barnes & Noble retail store (except for purchases made by check less than 7 days prior to the date of return) or (ii) 14 days of delivery date for Barnes & Noble.com purchases (except for purchases made via PayPal). A store credit for the purchase price will be issued for (i) purchases made by check less than 7 days prior to the date of return, (ii) when a gift receipt is presented within 60 days of purchase, (iii) textbooks returned with a receipt within 14 days of purchase, or (iv) original purchase was made through Barnes & Noble.com via PayPal. Opened music/DVDs/ audio may not be returned, but can be exchanged only for the same title if defective.

<u>After 14 days or without a sales receipt,</u> returns or exchanges will not be permitted.

Magazines, newspapers, and used books are not returnable. *Product not carried by Barnes & Noble or Barnes & Noble.com will not be accepted for return.*

Policy on receipt may appear in two sections.

Return Policy

<u>With a sales receipt, a full refund in the original form of payment</u> will be issued from any Barnes & Noble store for returns of new and unread books (except textbooks) and unopened music/DVDs/audio made within (i) 14 days of purchase from a Barnes & Noble retail store (except for purchases made by check less than 7 days prior to the date of return) or (ii) 14 days of delivery date for Barnes & Noble.com purchases (except for purchases made via PayPal). A store credit for the purchase price will be issued for (i) purchases made by check less than 7 days prior to the date of return, (ii) when a gift receipt is presented within 60 days of purchase, (iii) textbooks returned with a receipt

Looking into eyes

THE EYES REALLY ARE A WINDOW into the soul, or at least personality.

New research shows that the structure of the iris gives clues to personality. People with dense structures are warmer and more trusting, as well as more likely to be happier, than those with an open structure.

Men and women who have more furrows in their iris are more impulsive, neurotic and self-disciplined. People with a lot of contraction furrows were also less able to control their cravings and urges.

Those with denser irises have more positive emotions, and are more trusting and straightforward.

"We found that tissue differences in the iris, crypts and contraction furrows, are associated with personality. It gives real meaning to the expression that it is all in the eyes," says Mats Larsson, a behavioural scientist who led the study. "Our results suggest that tissue differences in the iris indeed can be used as a biomarker that reflect genetic and personality differences among people."

It has long been claimed that the eye may have links with personality, but most of the research has looked at eye colours. Some have found links, especially in children, but others have failed to confirm the findings.

The new research is based on the structure of the iris rather than the colour. It's based on the idea that genes involved in the development of the iris are also involved in the development of the part of the brain involved in personality.

The research, which also shows that up to 90 per cent of the tissue differences in the iris are due to genetic effects, is the first to look at the structure of the iris and personality.

In the study, reported in the journal *Biological Psychology*, the scientists at Örebro University and the Karolinska Institutet in Sweden analysed the eyes of more than four hundred people, as well as carrying out personality tests.

They looked in particular at two features of the iris, crypts and contraction furrows, which are related to thickness and density.

The significance of these features is that genes that influence the growth of the cells in the embryo are also involved in the developing of parts of the frontal lobe of the brain.

One gene in particular, Pax6, is involved in both. Brain researchers have shown that people with a mutation in the gene have higher rates of unusual behaviour, including impulsivity, and poor social skills.

Results showed that people with dense structures were more receptive to their inner feelings and tended to sympathize and feel concern for other people's needs more than people with less dense crypt features.

Men and women with dense structures form warmer and more trustful attachments to others and experience and express positive emotions, such as joy, happiness and excitement, more often than those with open crypt structures.

Contraction furrows, which relate to the thickness and density differences in all five cell layers in the iris, were also associated with personality.

People with many contraction furrows were less able than people with few contraction furrows to control their cravings and urges.

"Our results suggest that people with different iris features tend to develop along different personality lines," says Mats Larsson.

"If you have a dense iris structure you score higher on warmth than if you have an open iris structure. You have more positive emotions, more openness to experience, and more trust, and you are more straightforward and tender minded. The more crypts you have the lower score you have on warmth."

Comparing baby faces

NEWBORN BABIES LOOK MORE LIKE their mums than their dads.

Mums may say they look like their dads, but that's just a strategy women may have evolved over time to allay male fears about paternity.

And the more questionable the paternity, the greater the likelihood that the new baby will be said to look like Dad.

And it works, according to research, because once Dad sees himself in the baby, it makes him fatherly and more prepared to look after Mum and child.

"We found that mothers claim a paternal resemblance at birth that does not correspond to the actual resemblance, suggesting possible manipulation of the perception of facial resemblance to increase confidence of paternity," says Dr Charlotte Faurie of the University of Sheffield.

The research involved nearly one hundred babies, their parents and 260 independent judges, and the aim was to see whether there were any similarities between babies and young children and mums and dads. The view of mums and dads and those of the judges were then compared.

The results showed that at birth 100 per cent of mothers said boys looked like their father, and 77 per cent said girls looked like him. More than eight out of ten men thought the child looked like them.

But the independent judges found the opposite, with 50 per cent believing the babies looked like their mothers, and only a third finding any similarity with Dad.

"Fathers assigned resemblance of newborns to themselves in 83 per cent of cases, the opposite to that assigned by non-related judges," says the report.

"At birth, facial resemblance of a child ascribed by mothers is clearly biased toward the father.

"The evolution of this psychological mechanism to comfort fathers in their paternity by claiming paternal resemblance, allows women to eliminate subtle signals of deceit and is supported by the contradiction between what mothers claim about resemblance and actual resemblance assessed by external judges."

The report says that women have evolved counter-strategies that limit the problems of paternity uncertainty: "One is to ascribe the child's resemblance to the father. This is a way to ensure men of their paternity: unfaithful mothers could reduce uncertainty, whereas faithful mothers could gain more paternal investment."

It adds, "We found that mothers claim a paternal resemblance at birth that does not correspond to the actual resemblance, suggesting possible manipulation of the perception of facial resemblance to increase confidence of paternity."

Spotting liars

LIARS ARE NOT NERVOUS AND FIDGETY AFTER ALL.
They touch their nose, fiddle with their hair, and point less than people telling the truth, according to researchers. But they do move their hands up and down more, and use them to more to outline shapes of things they are talking about.

The research, which undermines the so-called Pinocchio theory of lying – that people scratch their nose more frequently when lying – shows that hand movements and gestures can be used to tell who is telling the truth.

The researchers warn that the results show that common techniques used by the police to spot those lying are unreliable.

Research also shows that people suspected of crimes are more likely to stay still during police interviews because they need to concentrate more than truth tellers.

"There is a popular perception that things like scratching the nose, playing with the hair, what we call self-manipulation gestures increase with people lying. People expect liars to be nervous and to be shifty and fidget more, but our research shows that is not the case," says Dr Samantha Mann, a psychologist at the University of Portsmouth, whose research with police suspects also shows that liars do not avoid eye contact.

"One theory is that people who are lying have to think harder, and when we think harder we tend to be a lot stiller, with fewer movements, because we are concentrating harder. Another possible explanation is that people expected liars to be nervous and make movements like touching their nose, so now, they are refraining from making those movements."

In the research, reported in the *Journal of Nonverbal Behavior*, the academics from Portsmouth and universities in Italy looked for changes in seven different types of hand movements – deictic, metaphoric, self-adaptor, rhythmic, emblematic, cohesive gestures and iconic – in 130 men and women when they were lying and when they were telling the truth. They also looked for any changes once they were told they were under suspicion.

The results show that liars use self-adaptor gestures – touching the nose or hair or other parts of the body – less than truth tellers. They use such gestures 15 to 20 per cent less. They also use deictic gestures – pointing – around 20 per cent less.

Metaphoric gestures – where the hands draw metaphoric shapes, such as a fist to show strength or a cup to show sums of knowledge – occurred 25 per cent more often when the participant was lying: "They are more frequently used when trying to make a convincing impression," say the researchers.

Emblematic gestures – movements that are familiar, like thumbs up or down, or a ring for OK – are also used slightly more often by liars, as are rhythmic gestures – moving the hand or finger up or down to the rhythm of what is being said.

Use of cohesive gestures – hand or finger movements that are unique to the individual, which may occur many times – are used slightly more by truth tellers. Iconic gestures – where the hands drawn real shapes in the air to illustrate what they are talking about, such as a clock or a house – is about the same.

The results show that when people were told they were under suspicion, they did change their hand movements, but the changes were similar in both groups: "Judging behavioural adaptations as signs of deceit or truth telling after suspicions are raised, a technique commonly used by the police, is therefore an unreliable method to detect deceit," say the researchers.

Funny genes

A BRITISH SENSE OF HUMOUR may be inherited, according to research on more than four thousand twins.

It shows that humour regarded as typically British has a genetic element in UK men and women, but not in Americans. While a genetic element was found to be involved in positive humour – telling jokes and looking on the bright side of life – in both countries, only in the UK were there links with negative humour – sarcasm and teasing.

Researchers say it may explain why the British like aggressively sarcastic, denigrating and self-deprecating humour in programmes such as *Fawlty Towers*, *Blackadder* and *The Office*.

"It is possible that differences exist between these nations in their sense of humour and that these may have led to different patterns of genetic and environmental influences," says Dr Rod Martin, one of the researchers.

"People in the UK, for example, may have a greater tolerance for a wide range of expressions of humour, including what many North Americans might consider aggressively sarcastic or denigrating: like *Fawlty Towers* and *Blackadder*. In the North American version of *The Office* the lead character is much less insensitive and intolerant than in the original UK version."

The researchers say that one possibility is that humour style is packaged with genes for personality.

In the research, reported in *Twin Research and Human Genetics*, the journal of the International Society for Twin Studies, the researchers looked at genetic and environmental contributions to humour style in nearly two thousand pairs of UK twins. In a second study, due to be published shortly in the journal *Personality and Individual Differences*, they looked at humour in 500 sets of North American twins.

The twins in the study, aged 18 to 74, completed a special questionnaire designed to be used to measure and classify sense of humour. The 32-item questionnaire is designed to identify four types of humour – two positive styles, affiliative and self-enhancing, and two negative, aggressive and self-defeating.

Affiliative humour is defined as a tendency to say funny things, to tell jokes, and to engage in witty banter in order to amuse others, to facilitate relationships, and to reduce interpersonal tensions. Self-enhancing is maintaining a humorous outlook on life, being amused by the incongruities of life, and use of humour to cope with stress.

Aggressive humour is used for criticizing or manipulating others, as in sarcasm, teasing and ridicule, as well as in the use of potentially offensive, sexist or racist forms of humour. Self-defeating or self-disparaging humour is amusing others by doing or saying funny things at one's own expense.

The assumption in twins research is that the environmental influences are the same for each twin. But genetic influences are twice as high in identical twins because they share all their genes, while fraternal twins share only 50 per cent.

By comparing the replies from identical twins with those from fraternal twins, the researchers were able to home in on the genetic element.

In the American twins, the researchers found that individual differences in affiliative and self-enhancing humour styles were largely attributable to both genetic and non-shared environmental factors. Differences in aggressive and self-defeating humour styles were largely attributable to shared and non-shared environmental factors.

In the UK data, from the Twin Research & Genetic Epidemiology Unit, King's College London, and St Thomas' Hospital, all four humour styles showed evidence of genetic influence. Identical twins were more than twice as likely to have the same kind of humour as fraternal twins.

"The purpose of the UK study was to attempt to replicate the North American finding that individual differences in affiliative and self-enhancing humour were largely attributable to additive genetic and nonshared environmental factors whereas individual differences in aggressive and self-defeating humour were largely attributable to shared and nonshared environmental factors," say the researchers. "But we found in the UK that all four styles of humour were best fit by an additive genetic and nonshared environmental model."

Dr Martin, from the Department of Psychology, University of Western Ontario, said the whole idea of the studies was to find out whether humour has a genetic basis.

"Most people think they have a good sense of humour, but when you probe and try to find out what they mean, then you find everyone has their own idea of what it is," he said. "Humour styles questionnaires measure the way you use humour, and individual differences in how we use humour in everyday life.

"In the North American sample we found that the two positive styles had a part genetic basis, while for the two negative styles, there was no evidence of any genetic basis.

"In the North American families, it seems to have been entirely learned. A shared environment was important, suggesting that growing up in a family where that kind of negative humour was practiced was important in the developing of that sense of humour.

"Then, in the UK, it turned out that all four styles had a genetic basis in the sample. Genetic basis to that kind of negative humour in the UK was close to 50 per cent."

Checkmate

WOMEN AREN'T BAD CHESS PLAYERS, they are just told they are.

Their reported underperformance is not down to lack of skills, but an awareness that they are expected to do badly in what's seen as a male bastion, according to new research.

In a series of Internet chess experiments, scientists have shown exactly how much effect gender stereotyping has on chess players. They showed that the performance of women drops by 50 per cent when they are playing a man. Yet, when they were told that the same opponent was a woman, they played as well as the man.

"In this paper it is argued that gender stereotypes are mainly responsible for the underperformance of women in chess. Gender stereotypes can have a greatly debilitating effect on female players leading to a 50 per cent performance decline when playing against males. Interestingly, this disadvantage is completely removed when players are led to believe that they are playing against a woman," say the researchers, whose report, "Checkmate? The role of gender stereotypes in the ultimate intellectual sport", is published in the *European Journal of Social Psychology*.

The researchers say that women are surprisingly underrepresented in the chess world, making up less than 5 per cent of registered tournament players worldwide and only 1 per cent of the world's grandmasters. Many reasons have been put forward, including lack of spatial skills in women, and greater aggressiveness and methodical ability in men, as well as more of a desire to win.

In the research, the academics looked at what happened when players did and did not know the sex of their opponent. The researchers were able to achieve that anonymity with chess games played over the Internet where the opponent is unseen.

Forty-two male–female pairs, matched for ability, played two chess games. In one part of the experiment, the players were unaware of the gender of their opponent and played against same-sex and opposite-sex opponents. They were also made aware of the stereotype that females are poor chess players.

The results show that when players were unaware of the sex of their opponent, the women played as well as the males. But when they were aware that they were playing against a male opponent, there was a dramatic drop in performance. When they falsely believed they were playing against a woman, they performed as well as their male opponents.

"Performance was reduced by about 50% when women were reminded of the stereotype and when they were aware of the fact that they were playing against a male opponent. In this case, they won only one fourth of the games. Yet, these same women were able to win half of the games when they were misled into believing that the opponent was female. The difference in performance is particularly impressive, if one considers that the opponent was exactly the same," say the researchers from the University of Padova, Italy.

"The difficulty encountered by female chess players may mainly reside in their awareness that others expect them to perform poorly in a predominantly male domain. Not only are females often accused of inferior ('girl's') play but, when they perform exceptionally well, their femininity is also often doubted.

"Our findings also suggest that, independent of experimental condition, women tend to approach chess games more cautiously and with lesser self-confidence, possibly because they are stigmatized in a male-dominated field. A second and more general message of our study is that self-confidence and a win-oriented promotion motivation contribute positively to chess performance. Since women show lower chess-specific self-esteem and a more cautious regulatory focus than males, possibly as a consequence of widely held gender stereotypes, this may at least in part explain their worldwide underrepresentation and underperformance in chess."

They add, "Thus, women seem disadvantaged not because they are lacking cognitive or spatial abilities, but because they approach chess competitions with less confidence and with a more cautious attitude than their male opponents."

Handedness

PEOPLE WHO ARE ABLE TO USE both hands with equal skill may not be so clever after all.

In fact, research suggests they may not be as bright as those restricted to being left- or right-handed.

"Our results provide support for the finding that there is a dip in intellectual performance among the ambidextrous relative to left- and right-handers," say the researchers, who found no difference between left- and right-handers.

Links between handedness and intellectual ability have long fascinated researchers, and in the new study psychologists analysed the replies to IQ tests, which also included information on handedness, of around 1,300 men and women.

Results showed that the IQ ranged from 85 to 135, with an average of 99.6. The average for the ambidextrous group was lower, however.

When researchers analysed different elements of the tests they found that the ambidextrous performed less well in arithmetic, spatial skills, language, memory and reasoning. Men who were ambidextrous scored higher than average in the category of social knowledge.

"Ambidexters scored lower than left- or right-handers on all tests except for social knowledge, where male ambidexters scored slightly higher," says the researchers, whose report appears in the journal *Neuropsychologia*. "Ambidextrous individuals perform more poorly than left- or right-handers, especially on subscales measuring arithmetic, memory, and reasoning."

Just why ambidexters are more at risk of deficits in intellectual performance is not clear. There has been a popular assumption that the ability to use both hands with equal skill is a sign of greater, not lesser, intelligence.

"One possibility is that reporting oneself to write with either hand is itself a manifestation of intellectual confusion, perhaps over which hand is which, or which hand is the preferred one," say the researchers from the University of Auckland.

Another possibility is that ambidexterity is a sign that one side of the brain is not dominant, as it is in most people, which leads to confusion over left and right.

The researchers point out that the lack of cerebral dominance has been associated with characteristics other than academic ability. There is some evidence, they say, of a susceptibility to illusions and paranormal thought.

"While these attributes may have negative connotations, they may also suggest creativity and perhaps charismatic leadership," says the report.

Chewing-gum watchers

WATCHING WHAT HAPPENS TO CHEWING GUM when it gets trodden on to pavements and into carpets may not be the most exciting research, but it has big implications.

In the UK alone, irresponsible gum cuds cost local councils almost £200 million a year to clean up, so anyone who comes up with a nonstick product would be on to a winner. But for more than twenty-five years, such a gum had proved impossible to develop.

The problem is that modern gums are made from synthetic latex to which softeners, sweeteners and flavourings are added. Such synthetic rubbers are stretchy, retain their properties indefinitely under all weather conditions, are resistant to aggressive chemicals and have strong adhesive properties, which is why they stick to just about anything.

A change in the stickiness of the gum would require a change in the chemical structure of the rubber gum base. But the gum base also determines commercially important and sensitive features of chewing gum such as the flavour, chewiness and freshness. The challenge that has proved so difficult for so long has been to develop a non-sticky or biodegradable gum base that does not compromise commercially critical requirements.

For chewing gum the balance between cohesion and adhesion is paramount in developing an easily removable product. For example, with very high cohesion and low adhesion, a discarded cud could just bounce off the pavement (like a rubber ball), but the gum would not be acceptable to chew. Achieving a balance between what is acceptable as a chewing gum and what is also easy to remove from surfaces is a delicate chemical and formulation problem.

To meet that challenge, researchers at Bristol University and a spin-out company, Revolymer, have been working on different recipes.

They produced more than two hundred different formulations to try to identify a chewing gum that is non-sticky; and in excess of a thousand pieces of gum have been made in a food science laboratory.

In various experiments, the researchers have stuck chewed gum to a variety of surfaces, as well as chewing it for several hours to see whether or not it loses its flavour or chewiness. The gum was stuck to glass, clothes, carpets, hair, shoes and just about anything else to find the best overall products.

Eventually, the researchers added a polymer that seemed to make the gum degrade and lose its stickiness in water. The polymer is incorporated into the chewing gum, replacing certain other non-ingestible and tacky ingredients, which creates a product with "good mouth feel" that is also easily removed from surfaces by water jets and scrubbers.

A number of street trials were carried out in towns in the UK with local councils to see how easily the new product could be removed compared with commercial gums. In trials, leading commercial gums remained stuck to the pavements three out of four times, but the new gum was removed within 24 hours by natural events, such as rain.

"The advantage is that it has a great taste, it is easy to remove and has the potential to be environmentally degradable," says Professor Terence Cosgrove, of the University of Bristol and Chief Scientific Officer of Revolymer. "The basis of our technology is to add an amphiphilic polymer to a modified chewing gum formulation which alters the interfacial properties of the discarded gum cuds, making them less adhesive to most common surfaces."

Expert blind tests of the product in tandem with commercial chewing gums showed it to have as good a chew in terms of mouth feel and texture as the leading brands. It can also be chewed for hours while keeping its pleasant soft chew.

Sales of chewing gum have been increasing steadily in recent years, with sugar-free gum the fastest-growing sector. Wrigley's, which owns many of the chewing-gum brands sold in the UK, has seen its sales across Europe and the USA grow by over a third since 1998, according to a report from the Department for Environment, Food and Rural Affairs (Defra).

There are various methods available for removing chewing gum from pavements.

Specialist gum removal companies typically charge between £0.45 and £1.50 per square metre, with the cost depending on the method, the type of surface and the amount of chewing gum: Trafalgar Square was cleaned in June 2003 at a cost of £8,500.

A national survey commissioned by Defra reported that chewing gum was the major source of staining on pavements and that the greatest problems are around schools, cinemas and swimming pools.

MEASURING FINGERS, LEGS, BREAST BOUNCE AND MARITAL HARMONY

Measuring compatibility

PERSONAL HYGIENE IS MORE IMPORTANT to a relationship than sex and physical attractiveness.

Tidiness is also more important than politics, while a shared liking for TV soaps carries the same weight as job status and sexual experience.

Research on British married couples shows that their compatibility can be calculated by comparing their answers to just 25 questions. Researchers who used the five-minute test on a speed-dating trial found they were able to accurately predict who would carrying on seeing each other and who would never see each other again.

"We found that when you got two people together for three minutes and then ask them whether they wanted to date, or be a friend, or never see the other person again, the test could very significantly predict the outcome," says Dr Glenn Wilson, reader in personality at the Institute of Psychiatry and lead author of a report on the research in *Sexual and Relationship Therapy*.

In the research, involving more than a thousand men and women, some two hundred married couples located from the UK electoral register were asked to complete detailed questionnaires about their marriages. They were also quizzed in a second test about the happiness of their relationship, and were invited to rate the relevant importance of the individual items in the questionnaire. Another 615 people were then also asked to rank the items.

For each of the 25 items, there are five possible answers. The difference between the responses are then scored, and the results analysed to see how comparable they are.

Scores above 100 indicate higher than average compatibility and scores below 100 suggest less than normal compatibility. The range is from 65 to 145.

The results show that for both sexes, the top five items were sexual fidelity, type of relationship sought, personal hygiene, desire to have children, and libido.

Personal hygiene compatibility scores were based on five possible answers to the question "How often would you take a bath or shower? – Twice a day or more, Sometimes twice a day, Once a day, About every other day, or Once a week or less".

Music preferences, height, belief in astrology, taste in food, and sexual experience were ranked as having the lowest importance to the relationship.

"Each of the questions has five response options. You can use it with couples who have been married for 30 years, and you can do it with people who have never met each other," says Dr Wilson.

"Because it is about compatibility it can only refer to the degree of similarity between a particular pair of people. You can use it to screen for partners, although when you meet they may not be your type.

"In a second paper we are preparing, we tested the idea on speed dating. We found that we could predict the outcomes for speed dating. We found that those who wanted to date scored 10 points higher than those who never wanted to see each other again. They scored 110 compared with 99.

"That is a very significant difference. The average score for a married couple is 116, and the score for a random couple is 100. So those speed daters who fancied each other were well on the way to the average score for a married couple.

"There are five responses to each item. It looks at the disparity and uses a complex formula to calculate a score. With fidelity, for example, if one person says they want an exclusive relationship, while the other wants a swinging arrangement, there would be a score of minus four.

"The formula is then used by which we arrive at a score for an average two people chosen at random of 100. The higher the score, the greater the compatibility."

Measuring marital disharmony

HAPPINESS IN A MARRIAGE PEAKS at the end of the first year.

And in marriages that end in divorce, the greatest levels of happiness in the relationship come a year before the wedding.

New research, the first big study to look at marriage, happiness and life satisfaction, also found that couples were happiest in marriages where one partner was a wage earner and the other a homemaker.

Married people were 30 per cent happier than singles, while cohabiting couples were 20 per cent happier than non-dating singles. Marriages where one partner stayed at home when they had children were also happier in the earlier years of the relationship.

The research also found that single people who get married are happier than those who remain single, and those who marry young are happier.

"We found evidence that happier single people are more likely to opt for marriage. Division of labour seems to contribute to spouses' well-being, especially for women and when there is a young family to raise. Spouses practising the division of labour report on average higher life satisfaction than dual income couples.

"In contrast, large differences in the partners' educational level have a negative effect on experienced life satisfaction," say the researchers.

In the research, the academics from the University of Zurich and the Institute for the Study of Labour in Bonn looked at seventeen years of data on a national German panel of men and women. Each year they were quizzed about how satisfied they were with life, and invited to score their happiness on a scale of 1 to 10. The researchers then compared the satisfaction rates in each year with the marital status of the men and women.

When they charted happiness over the seventeen years, they found that as the year of marriage approached, men and women had higher satisfaction scores. Soon after marriage, satisfaction declined.

The data show that the happiness score peaks at around twelve months into the marriage, with a second, smaller peak after five years.

Just why is not clear. One theory is that the transition to marriage causes changes in well-being. Another is that people adjust to the pleasant and unpleasant aspects of living with a partner in a close relationship.

The results also show that those who eventually divorce were always less happy than those who did not, even five years before marriage. The peak in happiness was one year before marriage. At the time of their marriage they were only marginally happier than they had been five years earlier.

The results also show that couples specializing after marriage – where one partner is in the labour market and the other is not – are better off in terms of life satisfaction than dual-income couples for the first eight years of a marriage.

"Women who, after marriage, live in households with complete division of labour report, on average, much higher life satisfaction scores," say the researchers.

"We find that spouses with small differences in their level of education gain, on average, more satisfaction from marriage than spouses with large differences. This sheds light on an aspect often neglected in the economic analysis of marriage: companionship. The enjoyment of joint activities or the absence of loneliness and the emotional support that fosters self-esteem and mastery are all important non-instrumental aspects contributing to the individual well-being of married people."

Measuring legs

LONG-LEGGED WOMEN REALLY ARE THE most attractive.

According to researchers the perfect shape for a woman is with legs that are 1.4 times the length of her upper body. But in men, shorter legs, the same length as the torso, are the most alluring.

Academics who have been measuring legs believe the attraction of long limbs in women and shorter versions in men may have evolved for different reasons.

Long legs, it seems, may be a sign of good health and good child-bearing capabilities in women, while short legs may make men look more muscular.

Although there has been some research on the attractiveness of height, the new study by psychologists from Liverpool University and University College London homed in on the ratio of leg and body length.

"A great many studies have considered height as an important component of physical attractiveness. A relatively unexplored approach is to consider the different components of height separately. One such component, recognized in clinical research but neglected otherwise, is the leg-to-body ratio," they say.

Although, on average, legs make up half of adult height, there are wide individual differences, and women tend to have higher leg-to-body ratios (LBRs).

In the study, reported in the journal *Body Image*, the researchers had volunteers rate male and female line drawings which had been adjusted to give different LBRs.

The legs were measured as the distance between the bottom of the feet and the top of the pelvic region (above the hips and below the waist). The body was measured as the distance between the top of the head and the pelvic region.

All the men and women rated both male and female images, and the results showed that both sexes preferred a higher LBR for the female images, and lower ratios for the male images.

The results show that the preferred ratio for women was 1.4, and for men 1.0.

"The results of this investigation are consistent with the idea that the LBR plays a role in judgements of male and female physical attractiveness. Overall, both male and female participants showed a preference for higher LBRs in women and lower LBRs in men," say the researchers.

"The study highlights a previously neglected sexually dimorphic feature of the human form in judgements of physical attractiveness. It lends some support to the hypothesis that a higher LBR increases female attractiveness but decreases male attractiveness."

Just why long legs are attractive in women but not in men is unclear. One theory is that long legs are a sign of fitness.

Some research, say the psychologists, suggests that tall women have wider pelvises than shorter women, allowing easier births and higher-birthweight babies.

Other work shows that relatively long legs are a sign of good health: "Interruption of growth results in a relatively long torso and short legs. If the rate of growth is sufficiently slowed down due to nutritional deficiencies or psychological stress, the adult will have shorter legs relative to the trunk," says the report.

Longer relative leg length is also associated with a reduced risk of heart disease, cancer and diabetes, and lower blood pressure, better cardiovascular profiles and lower adult mortality.

But while that may be why an attraction for long-legged women has evolved, it does not explain the attraction of shorter legs in men.

Measuring breast bounce

BREAST BOUNCE MOMENTUM, A JOGGING PERIL for women with large bosoms, may soon be contained, thanks to research dedicated to measuring the ups and downs.

Researchers, who have calculated that vertical movement can exceed 70 mm during jogging, say the stresses and strains can be so great that bra straps can damage nerves, causing numbness in the little finger.

But now the researchers have developed an intelligent fabric with its own sensors which can be used to more accurately design bras for women so there is less bounce in their athletic activities.

Because the female breast contains no supportive muscle or bone, support is needed to reduce breast motion, particularly during physical activity: "Due to this limited internal anatomical support, females are usually encouraged to wear external support in the form of a brassiere to reduce breast motion, particularly during physical activity," say the researchers.

Although effective in limiting breast motion, a consequence of current brassiere design is that the brassiere straps bear much of the load generated by breast momentum during physical activity, say the researchers, who report their findings in the *Journal of Biomechanics*.

"As breast mass increases, breast bounce momentum also increases, placing large loads on the straps and, in turn, excessive pressure on the wearer's shoulders."

They say the excessive pressure can cause deep brassiere strap furrows and can affect the underlying nerves. The pressure on the nerves can, they say, lead to abnormal sensations in the little fingers, numbness, tingling and burning. There can also be strap-related pain.

"Apart from strap-related pain, many females, particularly large breasted women, are restricted from participating in physical activity due to exercise induced breast pain associated with excessive vertical breast displacement," they say.

In the research at the University of Wollongong, scientists designed a special fabric with sensors to detect the smallest of movements as women walked and jogged on a treadmill.

Polymer-coated sensors were then stuck to each woman's bra, using

Velcro and adhesive tape. Sensors were placed on the right side of the brassiere on the strap, at different distances from the nipple.

Wires extending from the ends of each sensor were linked to a transmitter pack, worn around the waist of each woman, which sent the data on stress and strain to a Bluetooth telemetry receiver. Infrared emitting diodes were placed directly above the wire connections to record change in sensor length in millimetres and additional diodes were placed on each subject's right nipple under the brassiere and on the sternal notch to enable calculation of vertical breast displacement independent of torso motion.

Results showed that vertical breast displacement during walking and running ranged from 11 to 25 mm and 43 to 68 mm, respectively.

"It was concluded that, although polymer-coated fabric sensors may exhibit a small response lag due to textile geometry changes, they were able to accurately and reliably represent changes in the amplitude of vertical breast displacement during treadmill gait," say the researchers.

Finger measuring

CHILDREN WITH LONG RING FINGERS are more likely to be hyperactive and have behavioural and social problems.

Those with long index fingers are more likely to be neurotic and sensitive.

Research based on primary-school children in England and Austria shows that finger ratios are a strong indicator of risk for a variety of behavioural problems. It comes in the wake of other research which shows that the ratio of index and ring fingers is an indicator of sporting prowess, aggression, number of sexual partners, autism and vulnerability to depression.*

At the heart of the research is the theory that these two digits are markers for hormonal exposure in the womb, a historical record of what went on in the womb in the critical first three months. In particular, they are a sign of the levels of hormones that shaped the early life of the foetus when the brain, heart and other organs were growing.

A relatively long ring finger is a sign that these organs were exposed to higher levels of the male hormone testosterone, while a relatively long index finger is a marker of oestrogen exposure.

In the new study, researchers in Austria and the UK measured the finger ratios in schoolchildren in Lancashire and Vienna, and then gave detailed questionnaires assessing behavioural problems to their parents. The questionnaire asked about emotional symptoms, conduct problems, hyperactivity, inattention, peer relationship problems and social behaviour.

The results showed that the parents' ratings of greater problems with aggression and conduct disturbance in children were associated with lower ratios. And the lower the ratio, the greater the likelihood of problems.

A low ratio also predicted conduct problems and hyperactivity in the boys, and pro-social behaviour in the girls.

"We had two main findings. High testosterone before birth as

---

*Digit ratio is the length of the index finger divided by the length of the ring finger. If they are equal in length the ratio is one. If the ring finger is longer the ratio is less than one. Men in general have a longer ring finger relative to their index finger.

indicated by digit ratio produces quite a lot of behavioural problems in terms of conduct, temper tantrums, bullying, fights with other children, and that kind of thing, and hyperactivity, being easily distracted. In general it applies to boys and girls.

"It also reduces social behaviour, the tendency not to be concerned about other children's feelings, and not being helpful if someone is hurt," says Professor John Manning, an evolutionary psychologist at the University of Central Lancashire.

"The second point is these are very strong relationships indeed. These are the strongest relationships we have yet found for behaviour traits. Running speeds are also very strongly related to digit ratio to the point where there is a pretty good chance of predicting who is going to win a race.

"You start with a high ratio, which is to say a little bit above one, and you go down, say, to 0.9. For every increment you get an effect. As you go down the ratio the hyperactivity goes up. It also shows that the more testosterone you are exposed to the less neurotic or sensitive you are."

Measuring brain power

GEORGE W. BUSH HAS THE LOWEST IQ of all but one of his 42 predecessors in the White House.

Only General Ulysses S. Grant, who was a partner in a financial firm that went bankrupt after he retired as president, had a lower IQ, according to new research.

Ronald Reagan, much ridiculed for his intellectual powers, is not only ranked as having a higher IQ than Bush, he outranks George Washington too.

George W. was also the president judged to be least open to new experiences – he scored zero compared with a high of 99 for Jefferson – and his intellectual brilliance is rated by the researchers at minus 0.7.

The president with the highest IQ was John Quincy Adams, a Harvard lawyer who was the first president who was the son of a president.

"Bush's intellect may be more a liability than an asset with respect to his performance as the nation's chief executive. His strengths most likely lie elsewhere," say the researchers, whose study was published in the journal *Political Psychology*.

But the researchers say that he is still more intelligent that the average person: "Bush is definitely intelligent. The IQ estimates range between 111.1 and 138.5, with an average around 125. That places him in the upper range of college graduates in raw intellect. In addition, the figure is more than one standard deviation above the population mean, placing Bush in the upper 10% of the intelligence distribution," say the report's authors.

In the study, researchers calculated the IQ, leadership quality, openness to experience and intellectual brilliance of all the US presidents up to Bush himself.

Because most presidents of the United States died long before the advent of intelligence tests, the researchers used a range of sources to produce their estimates, including reviewing biographies and personality profiles, and using early childhood and adolescent accomplishments as part of the IQ estimate.

Personality descriptions were taken from biographies, identifying information removed, and independent judges asked to review them and rate presidents. Each president's pre-election publication record was reviewed, and IQs at different ages calculated. Data on five personality traits was used, including openness to experiences, such as new ideas, which is positively associated with intelligence.

Results show that John Quincy Adams tops the IQ league with 175, followed by Jefferson and Madison (160), Kennedy (159.8), Clinton (159), Carter (156.8), Adams (155), Arthur (152.3), Garfield (152.3) and Roosevelt (150.5).

The bottom ten are Washington (140), Harding (139.9), Taylor and Truman (139.8), A. Johnson (139.8), Buchanan (139.6), Taft (139.5), Monroe (138.6), George W. Bush (138.5) and Grant (130).

For intellectual brilliance, Jefferson comes top with a score of 3.1, followed by Kennedy with 1.8 and Wilson with 1.3. Harding and Coolidge come bottom.

For openness to experience, Jefferson. Lincoln and John Quincy Adams score over 90, ahead of Kennedy and Clinton. Bush scores the lowest – zero.

"Ever since George W. Bush was elected to the presidency, questions have emerged about his general intelligence. These results provide a more objective and quantitative means to address this issue," says the report's author, Dr Dean Keith Simonton of the University of California at Davis.

"Bush's IQ is below average relative to that subset of US citizens who also managed to work their way into the White House. In fact, his intellect falls near the bottom of the distribution. When compared with twentieth-century presidents from Theodore Roosevelt through Clinton, only Harding has a lower score.

"Bush's IQ falls about 20 points below that of his predecessor, Clinton, a disparity that may have created a contrast effect that made any intellectual weaknesses all the more salient," say the researchers. "Clinton's intellectual attainments as a Rhodes Scholar and Yale Law School graduate, his demonstrated capacity for mastering impressive amounts of complex and detailed information, his verbal eloquence and fluency, and his logical adroitness and sophistication – at times, as during the Monica Lewinsky scandal, verging on sophistry – place Clinton head and shoulders above his successor in terms of intellectual power."

All the signs suggest that Barack Obama, who served as a professor in the Law School at Chicago University from 1992 until his election to the US Senate in 2004, is likely to figure higher up the league table than his predecessor.

Sizing up women

MEN MAY BE ATTRACTED TO GOOD-LOOKING WOMEN BECAUSE they are more fertile.

Scientists have discovered that symmetrical women, those whose left sides are pretty much identical to their right, are significantly more fertile than other women.

The researchers found that non-symmetrical women had 30 per cent lower levels of female hormones, reducing the chances of conception.

"Our results suggest that, in women, symmetry is related to higher levels of oestradiol and, thus, higher potential fertility. As a consequence, men attracted to more symmetrical women may achieve higher reproductive success," say the researchers.

Symmetry is known to play a major part in attractiveness, and both men and women who are more symmetrical are considered to be more attractive.

But why symmetrical people are considered to be more attractive has been puzzling researchers. Research is now showing that there may be evolutionary reasons for being attracted to someone whose feet, ankles, hands, fingers, eyes, breasts, arms and ears are the same size and shape on each side.

It's suggested that such symmetry is a visible marker of good health, good genes and high fertility.

To test the theory, researchers at Harvard University and a number of other centres in Norway and Poland carried out detailed investigations on around two hundred women aged 24 to 36 with regular menstrual cycles, no fertility problem, and who were not using hormonal contraception.

The researchers used measurements of the fingers on the left and right hands of the women as markers of symmetry. The second and fourth digits of each hand were measured to the nearest millimetre. If the fourth left and fourth right fingers differed in length by up to 1 mm, the woman was classified as symmetrical, while those with differences greater than 2 mm were classed as asymmetrical.

Saliva samples were taken and tested for female hormones. The results show that there were considerable differences between the two groups of women. Symmetrical women had a 21 per cent higher mid-cycle oestradiol level than asymmetrical women. At other times the differences was as high as 28 per cent.

"Oestradiol produced during the menstrual cycle is crucial for successful conception, and levels of oestradiol are important indicators of a woman's ability to conceive," say the researchers.

They say that the average oestradiol levels are associated with the probability of conception of 12 per cent, while a 37 per cent rise in levels increases the probability to 35 per cent.

"In our study, oestradiol levels in symmetrical women were almost 30 per cent higher than those in asymmetrical women. Such difference in hormone levels suggests a substantial increase in the probability of conception for symmetrical women," they say.

The academics suggest that higher levels of female hormones may also explain why symmetrical women are healthier, because the hormones are known to stimulate the immune system.

The downside for symmetrical women is that high levels of female hormones during the reproductive years may cause problems in the post-menopausal years because, say the authors, high lifetime levels of reproductive hormones are related to an increased risk of breast cancer and other malignancies.

Other research at the University of New Mexico shows that people with symmetrical faces are better able to fight infections. The common cold, asthma and flu are all more likely to be combated efficiently by those whose left side matches their right. They measured the facial features of around four hundred young people and compared them with health records over three years, and found that those with matching sides were more healthy.

Researchers at the University of Western Australia have also found a link between symmetry in men and sperm quality. They have measured ear length, wrist diameter, elbow diameter, ankle diameter, foot length and foot width, in men aged 18 to 35, and found that those with left- and right-side differences had poor sperm quality. The finding may explain why women are unconsciously attracted to symmetrical male faces.

Measuring sleep

SLEEP IS NOT JUST ABOUT DROPPING OFF into the land of Nod.

There are good and bad ways of doing sleep, and according to researchers, there is an entire social etiquette to sleep and sleeping.

There are, they say, socially appropriate and inappropriate, prescriptive and proscriptive ways of sleeping in every-night life.

"How we sleep, when we sleep, where we sleep, what we make of sleep, and with whom we sleep, are all socially, culturally and historically variable matters," say University of Warwick researchers whose report appears in *Sociology*.

"Sleep reveals a whole side of our life in which our waking selves, and the conscious controls and civilized codes of conduct upon which they are predicated, are relinquished on a daily or nightly basis. All sorts of weird and wonderful things can happen while we sleep: things which may prove a source of embarrassment, if not downright shame or humiliation."

The report says that the loss of waking consciousness that sleep involves does not make it a non-experience. In fact, in many cases the individual may only be half asleep.

"Murders indeed have been committed by people while asleep. Dieters too have been captured on camera making unknown trips to the kitchen in the night, gorging on those favourite foods that their waking conscious selves and dietary regimes prohibit," says the report.

According to research, there are a number of different types of sleep:

Sleep feigners: People feign sleep for various social ends. "Women, or men for that matter, may also feign sleep in order to avoid the unwanted or amorous attentions of their bed mates." They may also, for example, pretend to be asleep in order to listen in on someone else's conversation, or worse.

Socially attentive sleepers: Although sleep may involve a loss of waking consciousness, it is possible to be roused by someone calling your name. "Parents' sleep, particularly a mother's sleep, may also be significantly 'retuned', such that the slightest cry or stirrings of the newborn infant awakens them."

Snoozers: Also known as quasi-sleepers, they include commuters on trains who are able to sleep without missing their stop. "In doing so, moreover, they frequently remain within the norms of propriety in public places by sleeping without complete loss of bodily control, deference or demeanour, thereby according due respect to their fellow travellers."

Nappers: "No longer a furtive or fugitive act of rebellion, the nap is now becoming an acceptable if not officially prescribed part of the working day, within certain sectors or segments of the labour force at least. Some companies are building their own dedicated 'nap rooms' and 'nap tents' for employees."

Intolerant sleepers: These are the people who prioritize their own sleep to the detriment of significant others. "If I like to sleep with the light off and my partner likes to read in bed, then to the extent that her bedtime reading is curtailed or cut short by my lights out policy, then again I am open to the charge of being an inconsiderate, selfish or intolerant sleeper."

Selfish sleepers: As well as snorers and sleepwalkers who fail to seek treatment, this category includes those men who hear the baby crying and turn over. "If, for example, my partner needs comforting in the middle of the night for some reason, or it is my turn to get up and feed or comfort our waking children, then I may very well be construed as inconsiderate or selfish if I default on these nocturnal duties in favour of my own sleep."

Waist measuring

BRITISH MEN AND WOMEN ARE bursting out all over.

Men's waists now average more than 37 inches, and the typical female figure is now 39:34:41. A study based on body scans of nearly nine thousand UK adults has also found big differences in the average American and British body shapes.

The differences, say researchers, may explain why Americans have a higher risk of a number of diseases, including diabetes, high blood pressure, heart disease, stroke and cancer, than their UK counterparts.

And they say three-dimensional body scans could be a new, cheap way to identify the growing numbers of people at risk and how they respond to treatment.

"Our analyses have revealed significant differences in body shape between US and UK white adults. These differences may prove to play a key role in accounting for differences in morbidity and mortality," say the researchers from University College London, whose study appears in the *International Journal of Obesity*.

In the study, the researchers set out to compare differences in the shapes of British and American adults. They used data obtained on 3,907 men and 4,710 women, all aged over seventeen, from the UK, and a similar number from America. All had been measured with special scanning equipment.

Results showed that the average British woman had an average weight of 66 kg or 145 pounds, just over 10 stone, while the typical man weighed in at 80.3 kg or 177 pounds. American men and women were 8–13 pounds heavier.

A total of 38.8 per cent of the British men and 27.3 of the women were overweight; 13.7 per cent of both sexes were obese. Similar numbers of Americans were overweight, but 23.6 per cent of men and 21.3 per cent of white Americans were obese.

The average bust in the survey was 99 cm or 38.9 inches in the UK, and 103 cm or 40.5 inches in the USA.

The average British thigh in the survey was about the same for men and women, around 49 cm or 20 inches. There was only 1 cm difference in hips, with men measuring 103 cm and women 104, compared with 107 cm in American women.

Results also showed that waists have increased in size in both countries. Half a century ago, the average British woman's waist was reported to be 27 inches. According to the report, the average of the women surveyed was 87.4 cm or 34 inches. The American female waist was even bigger, at 88.4 cm.

"In both sexes, after adjusting for age and height, Americans had significantly greater weight, BMI and girths than their UK counterparts," says the report. "American white men had larger waists relative to physique than UK white men, whereas American white women had smaller waist girth than UK white women."

The researchers say the differences they found may help account for the differences in illness and mortality between the two countries.

They say substantial evidence now links the increasing prevalence of obesity with an increased risk of disease, but that the way obesity is measured – body mass index or BMI – is a poor indicator of fat levels because individuals differ in their distribution of excess weight.

"A major factor hindering investigation of this issue has been the use of a crude approach to categorizing obesity. BMI provides only a poor proxy for the central fat mass most strongly associated with disease risk," they say.

They add that the three-dimensional scanning of people in their underwear used in the surveys may be an accurate way of assessing risk.

"Our findings highlight the potential for 3D body scanning to contribute to the categorization of risk and the monitoring of patients," they say.

"Providing a wealth of information about body shape at a fraction of the cost of MRI scanning, 3D scans combined with ethnic-specific reference data have the potential to identify those at high risk of the metabolic syndrome, and to track the response of such individuals to treatment."

D'YA THINK STRIP LIGHTING'S SEXY – OR WOULD YOU RATHER WATCH CAR PAINT DRY?

OF COURSE THERE IS MORE, much more.

At this very moment, thousands of researchers are beavering away on projects that may seem bizarre, bewildering or just plain boring, but which all have a serious objective.

Finding out exactly when cockroaches prefer to mate (early to late evening, actually) may not seem to be going anywhere, but it could help develop new strategies for tackling the pests. And collecting and measuring cigarette butts from different neighbourhoods may seem to have little purpose, but it is helping doctors in New Zealand to identify the most intense smokers (butts from poor, deprived areas were the shortest).

A number of researchers are currently looking at how long it takes car paint to dry and metals to rust, while others are occupied with finding out how fluorescent lighting affects sexual behaviour, or investigating the courtship of newts.

Watching snails play follow-my-leader, probing the sexual behaviour of goats, and calculating the breaking strength of spaghetti are also under investigation. Projects measuring flatulence, watching bread rise and trying to get human eyelashes to grow an extra 2 mm are also work in progress.

Scientific research of all kinds is at record levels, in terms of numbers of practitioners and projects, fuelled by university expansions, the needs of thousands of scientific journals (the publisher Elsevier alone now has a database of over nine million papers) and a growing public thirst for knowledge (there has never been a better time for researchers).

In such a friendly climate, there is only one real obstacle – funding. But that is nothing new. As Einstein himself put it, "Science is a wonderful thing if one does not have to earn one's living at it."

REFERENCES

1 DIRTY NAPPIES, URINE, BODY ODOUR, SHIRT SMELLERS, AND 100,000 OLD TOENAILS
Looking into dirty nappies
Development of the Human Infant Intestinal Microbiota. *PLoS Biol* 5:
doi:10.1371/journal.pbio.0050177

How urine could save the world from famine
Pure human urine is a good fertilizer for cucumbers: *Bioresource
Technology*: 98: 214–217

Smelling body odour
Non-advertised does not mean concealed: body odour changes across the
human menstrual cycle: *Ethology*: 112: 81–90

Shirt smellers
The effect of meat consumption on body odor attractiveness: *Chemical
Senses*: doi:10.1093/chemse/bji017

Toenail clippings
Toenail nicotine levels as predictors of coronary heart disease among
women: *American Journal of Epidemiology*: doi:10.1093/aje/kwn061

Smelling fear
Chemosignals of fear enhance cognitive performance in humans: *Chemical
Senses*: doi:10.1093/chemse/bjj046

2 COUNTING KISSES, FAST FOOD SHOPS, STAIRS, YAWNS AND HOW MANY TIMES PEOPLE SAY SORRY

Fast food shops

Neighbourhood fast food environment and area deprivation – substitution or concentration?: *Appetite*: doi:10.1016/j.appet.2006.11.004

Stair counters

Modest effects of a controlled worksite environmental intervention on cardiovascular risk in office workers: *Preventive Medicine*: doi:10.1016/j. ypmed.2006.11.005

Counting kisses

Kissing laterality and handedness: *Laterality*: 11: 573–579

Counting offsides

Errors in judging "offside" in association football: test of the optical error versus the perceptual flash-lag hypothesis: *Sports Science*: 24: 521–528

Counting chores

National context and spouses' housework in 34 countries: *European Sociological Review*: doi:10.1093/esr/jcm037

How many times people say sorry

Social variation in the use of apology formulae in the British national corpus: *The Changing Face of Corpus Linguistics*. Edited by Antoinette Renouf and Andrew Kehoe: 205–221

Counting yawns on trains

Do men yawn more than women?: *Ethology and Sociobiology*: 10: 375–378

Totting up high heels

Women pay attention to shoe heels: *European Urology*: doi:10.1016/j. eururo.2008.01.046

3 WATCHING ANTS, COUNTING WORMS, LISTENING TO DOGS, AND GRUNTING AT BABOONS

Watching ants find their way home
Do leafcutter ants, Atta colombica, orient their path-integrated home vector with a magnetic compass?: *Animal Behaviour:* doi:10.1016/j. anbehav.2007.09.030

Measuring the genitalia of the polar bear
Spatial and temporal variation in size of polar bear (Ursus maritimus) sexual organs and its use in pollution and climate change studies: *Science of the Total Environment:* 387: 237–246

Overturning tortoises and the brain
Lateralized righting behavior in the tortoise (Testudo hermanni): *Behavioural Brain Research:* 173: 315–319

Counting worms
Earthworm community structure on five English golf courses: *Applied Soil Ecology:* doi:10.1016/j.apsoil.2008.02.001

Cow's teeth
Origin, paleoecology, and paleobiogeography of early Bovini: *Palaeogeography, Palaeoclimatology, Palaeoecology:* 248: 60–72

The sheep watchers
Sheep self-medicate when challenged with illness-inducing foods: *Animal Behaviour:* 71: 1131–1139

Plight of the bumblebee
Rarity and decline in bumblebees – A test of causes and correlates in the Irish fauna: *Biological Conservation:* doi:10.1016/j.biocon.2006.11.012

Listening to dogs barking
Dogs can discriminate barks from different situations: *Applied Animal Behaviour Science:* doi:10.1016/j.biocon.2006.11.012

Grunting at baboons
Baboons eavesdrop to deduce mating opportunities: *Animal Behaviour:* doi:10.1016/j.anbehav.2006.10.016

Avoiding crowds
A note on the influence of visitors on the behaviour and welfare of zoo-housed gorillas: *Applied Animal Behaviour Science*: 93: 13–17

Strewth, Joey, cover your eyes
Heterosexual and homosexual behaviour and vocalisations in captive female koalas (Phascolarctos cinereus): *Applied Animal Behaviour Science*: 103: 131–145

4 TALKING TO PENGUINS, SPOTTING FAT BIRDS, AND RUDDY DUCKS
Dawn chorus
Dawn song in superb fairy-wrens: a bird that seeks extra-pair copulations during the dawn chorus: *Animal Behaviour*: doi:10.1016/j.anbehav.2007.05.014

Counting birds
House sparrow (Passer domesticus) habitat use in urbanized landscapes: *Journal of Ornithology*: doi:10.1007/s10336–007–0165-x

Talking to penguins
Name that tune: call discrimination and individual recognition in Magellanic penguins: *Animal Behaviour*: 72: 1141–1148

Picking up penguins
Huddling behavior in emperor penguins: Dynamics of huddling: *Physiology & Behavior*: 88: 479–488

Who's a clever boy, then?
Cognitive and communicative abilities of Grey parrots: *Applied Animal Behaviour Science*: doi:10.1016/j.applanim.2006.04.005

Ruddy ducks
The ruddy duck Oxyura jamaicensis in Europe: natural colonization or human introduction?: *Molecular Ecology*: 15: 1441–1453

Who's a pretty (big) boy, then?
Preliminary assessment of the effect of diet and L-carnitine supplementation on lipoma size and bodyweight in budgerigars: *Journal of Avian Medicine and Surgery*: 18: 12–18

Spotting fat birds
Recent changes in body weight and wing length among some British passerine birds: *Oikos*: 112: 91–101

Sexual fetishism in a quail
An animal model of fetishism: *Behaviour Research and Therapy*: 42: 1421–1434

5 CLIFF RICHARD, TALKING TO WITCHES, THE QUEEN, AND BRITISH PRIDE
British pride
The decline of British national pride: *British Journal of Sociology*: 58: 4

Talking to witches
Consumer witchcraft: are teenage witches a creation of commercial interests?: *Journal of Beliefs & Values*: 28: 45–53

Cliff-hanger
Cliff Richard's self-presentation as a redeemer: *Popular Music*: 27/1: 77–97

Name games
Testing the generality of the name letter effect: name initials and everyday attitudes: *Personality and Social Psychology Bulletin*: 31: 1099–1111

Probing the brains of taxi drivers
London taxi drivers and bus drivers: a structural MRI and neuropsychological analysis: *Hippocampus*: 16: 1091–1101

Football strikers really are fit
Does the face reveal athletic flair? Positions in team sports and facial attractiveness: *Personality and Individual Differences*: doi:10.1016/j. paid.2007.05.020

We speak the Queen's English, we does
An acoustic analysis of "happy-tensing" in the Queen's Christmas broadcasts: *Journal of Phonetics*: doi:10.1016/j.wocn.2005.08.001

6 FETISHES, ONE-NIGHT STANDS, AND SIZE DOES MATTER
Making sex five times better
Why humans have sex: *Archives of Sexual Behavior*: doi:10.1007/ s10508-007-9175-2

Fetishes
Relative prevalence of different fetishes: *International Journal of Impotence Research*: 2007: 1–6

Time watchers
Sex differences in simple visual reaction time: a historical meta-analysis: *Sex Roles*: 54: 57–69

Making men fat
Weight halo effects: Individual differences in personality evaluations and perceived life success of men as a function of weight?: *Personality and Individual Differences*: 42: 317–324

Telling jokes
Production and appreciation of humor as sexually selected traits: *Evolution and Human Behavior*: 27: 121–130

Size does matter (all 5.3 inches)
Does size matter?: men's and women's views on penis size across the lifespan: *Psychology of Men and Masculinity*: 7: 129–143

Studying one-night stands
The morning after the night before: affective reactions to one-night stands among mated and unmated women and men: *Human Nature*: doi:10.1007/s12110–008–9036–2

Ear testing
Emotion words are remembered better in the left ear: *Laterality*: 10: 149–159

7 HOW TO BOIL AN EGG, USE DANDRUFF, AND WATCH CABBAGES GROW
Watching cabbages grow
Estimating cabbage physical parameters using remote sensing technology: *Crop Protection*: doi:10.1016/j.cropro.2007.04.01

Egg-boiling
The science of boiling an egg: http://newton.ex.ac.uk/teaching/CDHW/egg/

The dandruff detectives
Dandruff as a potential source of DNA in forensic casework: *Journal of Clinical Forensic Medicine*: 6: 58

Egg-cracking
Effect of egg shape index on mechanical properties of chicken eggs: *Journal of Food Engineering*: 85: 606–612

Why wallpaper never comes off in one piece
Tearing as a test for mechanical characterization of thin adhesive films: *Nature Materials*: 7: 386–390

Watching goalkeepers dive
Action bias among elite soccer goalkeepers: The case of penalty kicks: *Journal of Economic Psychology*: 28: 606–621

Why fat people are looked down on
Pathogen-avoidance mechanisms and the stigmatization of obese people: *Evolution and Human Behavior*: doi:10.1016/j.evolhumbehav.2007.05.008

Bad driving runs in families
Couple similarity for driving style: Transportation research Part F: *Traffic Psychology and Behaviour*: 9: 185–193

Watching the weather
Clouds make nerds look good: field evidence of the impact of incidental factors on decision making: *Journal of Behavioral Decision Making*: 20: 143–152

Weighing up climate change
Greenhouse gas benefits of fighting obesity: *Ecological Economics*: doi:10.1016/j.ecolecon.2007.09.004

8 HOW TO CALCULATE WINNERS, SPOT LIARS, AND STOP CHEWING GUM STICKING TO PAVEMENTS
Calculating winners
Combining player statistics to predict outcomes of tennis matches: *Journal of Management Mathematics*: 16(2): 113–120

Eyebrow testing
Attractiveness of eyebrow position and shape in females depends on the age of the beholder: *Aesthetic Plastic Surgery*: doi:10.1007/s00266-006-0149-x

Looking into eyes
Associations between iris characteristics and personality in adulthood: *Biological Psychology*: doi:10.1016/j.biopsycho.2007.01.007

Comparing baby faces
Differential facial resemblance of young children to their parents: who do children look like more?: *Evolution and Human Behavior*: doi:10.1016/j. evolhumbehav.2006.08.008

Spotting liars
The impact of deception and suspicion on different hand movements: *Journal of Nonverbal Behavior*: doi:10.1007/s10919-005-0001-z

Funny genes
Genetic and environmental contributions to humor styles: a replication study: *Twin Res Hum Genet*: 11(1): 44-47

Checkmate
Checkmate? The role of gender stereotypes in the ultimate intellectual sport: *European Journal of Social Psychology*: doi:10.1002/ejsp

Handedness
Handedness and intellectual achievement: An even-handed look: *Neuropsychologia*: doi:10.1016/j.neuropsychologia.2007.09.009

Chewing-gum watchers
Chewing gum watchers: www.revolymer.com

9 MEASURING FINGERS, LEGS, BREAST BOUNCE AND MARITAL HARMONY

Measuring compatibility

Measurement of partner compatibility: further validation and refinement of the CQ test: *Sexual and Relationship Therapy*: doi:10.1080/14681990500161723

Measuring marital disharmony
Does marriage make people happy, or do happy people get married?: *The Journal of Socio-Economics*: doi:10.1016/j.socec.2005.11.043

Measuring legs
The leg-to-body ratio as a human aesthetic criterion: *Body Image*: 3: 317–323

Measuring breast bounce
Can fabric sensors monitor breast motion?: *Journal of Biomechanics*: 40: 3056–3059

Finger measuring
The 2nd to 4th digit ratio and developmental psychopathology in school-aged children: *Personality and Individual Differences*: doi:10.1016/j.paid.2006.07.018

Measuring brain power
Presidential IQ, openness, intellectual brilliance, and leadership: estimates and correlations for 42 US chief executives: *Political Psychology*: 27: 511–526

Sizing up women
Symmetrical women have higher potential fertility: *Evolution and Human Behavior*: doi:10.1016/j.evolhumbehav.2006.01.001

Measuring sleep
The social etiquette of sleep: some sociological reflections and observations: *Sociology*: 41: 313–328

Waist measuring
Body shape in American and British adults: between-country and inter-ethnic comparisons: *International Journal of Obesity*: doi:10.1038/sj.ijo.0803685

ABOUT THE AUTHOR

ROGER DOBSON is an award-winning freelance journalist who con-tributes to a number of national newspapers on health and science, including the *Daily Mail*, *The Times* and *Sunday Times*, and the *Independent* and *Independent on Sunday*, as well as the *British Medical Journal*. He and his family live on the slopes of the Skirrid mountain near Abergavenny in Monmouthshire, South Wales. He is the author of *Death Can be Cured, and 99 other medical hypotheses*.

ALSO BY ROGER DOBSON

Ever wondered why babies suck their fingers, or why humans have chins? Want to know how bad TV shows cause dementia, or find out why women groan during sex? Or are you curious about how the moon causes gout attacks, and how shaving increases the risk of cancer? Or would you just like to know the date you will die?

Look no further than *Death Can Be Cured*. Over the years, hundreds of scientists, academics, doctors and independent researchers have come up with new answers, explanations and theories for almost everything, from AIDS and arthritis to vomiting and zinc. There are new theories to explain the Turin shroud, why religious revelations have always occurred on mountains, and why Queen Elizabeth I never married.

If you're curious about health, science, and the world around – and above – you, read on.

ISBN 978-1-905736-31-7/£8.99 Paperback